Analysis and Design of Steel and Composite Structures

Analysis and Design of Steel and Composite Structures

Editor

Harshad Bhandari

scitus
academics

Analysis and Design of Steel and Composite Structures

Edited by **Harshad Bhandari**

ISBN: 978-1-68117-003-9

Library of Congress Control Number: 2015931519

SCITUS Academics LLC,
www.scitusacademics.com
Box No. 4766, 616 Corporate Way,
Suite 2, Valley Cottage,
NY 10989

Notice

Printed in United States of America on Acid Free Paper ∞

Contents

Preface

This book, Analysis and Design of Steel and Composite Structures offers a comprehensive introduction to the analysis and design of both steel and composite structures.

Design of steel and composite structures is the design of compression members, effective lengths of columns, design of plate girders design by buckling analysis, design of portal frames, behaviour and design of beam-columns, connection design, plastic design (beams, simple frames), composite steel-concrete structures, elastic and rigid plastic analysis of composite beams, composite columns, composite connections.

Composite construction is the dominant form of construction for the multi-storey building sector. Its success is due to the strength and stiffness that can be achieved, with minimum use of materials.

The reason why composite construction is often so good can be expressed in one simple way - concrete is good in compression and steel is good in tension. By joining the two materials together structurally these strengths can be exploited to result in a highly efficient design. The reduced self weight of composite elements has a knock-on effect by reducing the forces in those elements supporting them, including the foundations. Composite systems also offer speed of construction benefits, which is a key reason for the boom in use of steel for commercial buildings. The floor depth reductions that can be achieved using composite construction can also provide significant benefits in terms of the costs of services and the building envelope.

Editor

Chapter 1

A Practical Data Recovery Technique for Long-Term Strain Monitoring of Mega Columns During Construction

Se Woon Choi, EunMi Kwon, Yousok Kim, Kappyo Hong, and Hyo Seon Park

Department of Architectural Engineering, Yonsei University, Seoul 110-732, Korea

ABSTRACT

A practical data recovery method is proposed for the strain data lost during the safety monitoring of mega columns. The analytical relations among the measured strains are derived to recover the data lost due to unexpected errors in long-term measurement during construction. The proposed technique is applied to recovery of axial strain data of a mega column in an irregular building structure during construction.

The axial strain monitoring using the wireless strain sensing system was carried out for one year and five months between 23 July 2010 and 22 February 2012. During the long-term strain sensing, three different types of measurement errors occurred. Using the recovery technique, the strain data that could not be measured at different intervals in the measurement were successfully recovered. It is confirmed that the problems that may occur during long-term wireless strain sensing of mega columns during construction could be resolved through the proposed recovery method.

INTRODUCTION

As the shapes of building structures become irregular or their height and size increase, mega columns, whose cross sectional area is larger than that of conventional members, are used more frequently. A mega column, a vertical member that comprises the resisting system against a lateral load such as earthquake or wind load, is placed at perimeter of a building plane in order to efficiently control the lateral displacement of the structure [1, 2].

For structures in which relatively large mega columns have been used, the safety of the members are managed during construction through structural health monitoring (SHM) using various sensors, including axial strain sensors [3–5]. Various sensing methods have been developed for field measurements of strains in buildings and infrastructures [6–10]. Recent significant developments of new strain sensors and their applications can be found in [11–14]. A wireless sensor network is preferred for the SHM of large-scale structures under construction because of the potential instability of the electricity supply and possibility of unexpected damage to the sensors and transceiver lines, and the real-time evaluation of the safety of structural members should be taken into account [15, 16].

However, it is difficult for a wireless sensor network system not to lose data measured by the sensors for various unexpected reasons such as communication problems, noise, failure or abnormal installation of the sensors, harsh weather conditions, and hardware failures [17, 18]. Various studies on the recovery of lost data have mostly been focused on the failure tolerance of the wireless sensor networks [19, 20]. The concept of large-scale neuron sensor networks or a carbon nanotubes

(CNT)-based artificial neural system (ANS) [11,12,21] or the use of artificial neural networks in management of strain data in a sensor network [22,23] may be used in estimation of strain data without recovery of abnormal sensors.

In this study, a practical data recovery method is proposed for the data loss that occurred during the safety monitoring of mega columns in a large-scale irregular structure that was actually under construction. The on-site safety monitoring of the constructed mega columns was accomplished by measuring the longitudinal strains with four wireless vibrating wire strain gauges (VWSG) installed vertically on the same cross section [15]. In this study, the analytical correlation among the longitudinal strains measured by the four wireless VWSG sensors was derived for the recovery of lost data. Based on the correlations, the strain data that could not be measured at each of the measurement intervals in the measurement data for a total of one year and five months between 23 July 2010 and 2 February 2012 were recovered, and the errors contained in the data measured by the damaged sensors were modified.

STRAIN SENSING SYSTEM FOR A MEGA COLUMN IN THE IRREGULAR BUILDING UNDER CONSTRUCTION

A practical monitoring system is applied to wireless strain sensing of the mega column in an irregular building (Figure 1) composed of three basement levels and four levels above the ground. As shown in Figure 1, a 26.1 m long mega column is used to support the mega truss. Since the structure is mainly used for exhibitions, large-scale structural members such as mega columns and mega trusses were employed to secure sufficiently large open spaces.

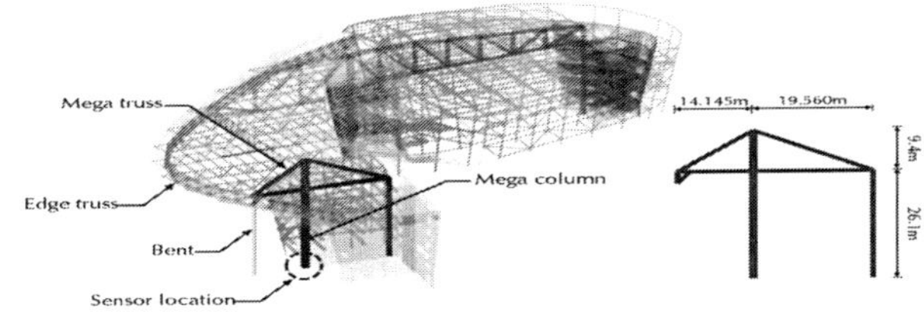

Figure 1:.Mega column in a large-scale irregular building structure.

Due to the advantages of simplicity in the operating principles and the low cost of installation compared to other sensors, VWSGs are the most frequently used sensors in the construction field and for static strain monitoring of building structures. In this study, Geokon's VSM-400 model [24] with a length of 150 mm as the VWSG was used, installed at the centers of the section of the mega column on the fourth basement level of the irregular building shown in Figure 1.

The measurement system for the sensing of longitudinal strain of the mega column consisted of four VWSGs installed on the mega column, a sensor node that transmits the information measured by the sensors to the master node through a wireless transmission system, and a master node that transmits all the measurement data received from the sensor node to the monitoring server through the long distance wireless transmission system shown in Figure 2.

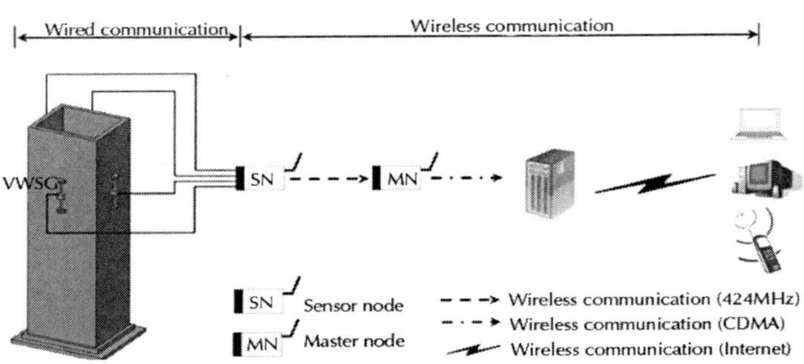

Figure 2: Wireless strain sensing system for the mega-column based on VWSG.

As shown in Figure 3a, the wireless sensor node for the 4-channel VWSG system consists of a sensor processing module, processor, memory, communication module, and power supply unit [15]. During construction, the sensor node inevitably requires batteries where continuous power supply through a wire is impossible. The low power consumption technology (LPCT) was adopted to resolve the power supply problem. The LPCT is realized by controlling the nodes through the active-sleep mode. In active mode, a sensor node performs every function. To minimize power consumption, sleep mode makes a sensor node inactive, which means that transmitting and receiving of the data is disabled. In addition, the transceiver and peripheral circuit are inserted into a single chip (CC1020 transceiver by Chipcon) which is used in communication with 424 MHz UHF within the Industrial Scientific and Medical (ISM) bandwidth, to operate with low power consumption [25]. In this study, the sensor node is designed to consume an electric current of 50 mA in active mode and 200 μA in sleep mode. In addition, the sensor nodes are also operated by external DC power.

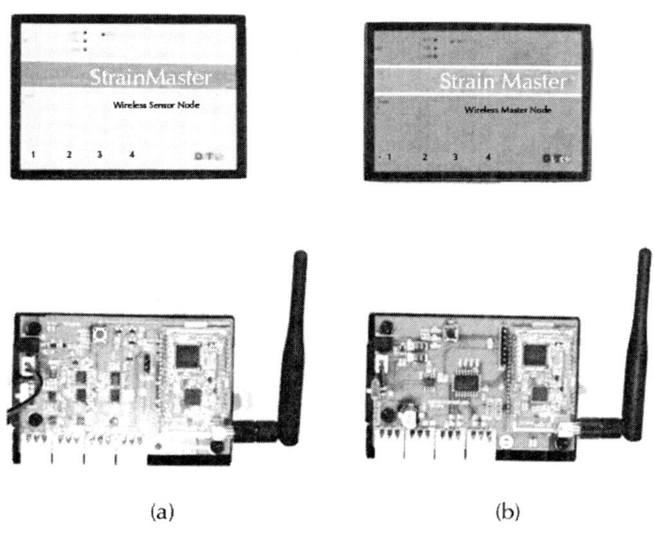

(a) (b)

Figure 3: Sensor node and master node for wireless sensing system. (a) Sensor node (b) Master node.

A master node (Figure 3b) serves the role of transmitting the data to the monitoring server located at a long distance by mobile communication after receiving the measured values from sensor nodes. A master node also has power issues. Thus, LPCT with active-sleep mode, which is already applied in the sensor nodes, was also adopted in the master node. If sensor nodes are in active mode, but the master node is in sleep mode, data cannot be transmitted to a monitoring server. Thus, the setting of active-sleep mode among sensor-master node should be in identical manner for successful transmission.

The data is transmitted between the master node and the monitoring server by some mobile communication technology. The master node consists of two parts; a receiver module and Code-Division Multiple Access (CDMA) transmitter. The receiver module of the master node is directly connected with a commercially available CDMA transmitter through a short cable. After the receiver module receives and delivers the measurement data from the sensor nodes to the cable, the data is transmitted at the CDMA transmitter by means of the mobile communication system, as shown in Figure 2.

ANALYTICAL RELATIONSHIP AMONG MEASURED LONGITUDINAL STRAINS

The longitudinal strain $_{i}$ (i = 1 to 4) measured at the four sensors of S1, S2, S3, and S4 installed on the cross section of a mega column in Figure 4 can be expressed as the sum of the strains generated by the axial force, 0, and the strains generated by the bending moment applied on the x and y axes, $_{Mx}$ and $_{My}$. Since the strain due to the bending moment is proportional to the distance from the neutral axis, a compression strain and a tensile strain of the same magnitude are generated simultaneously. In Figure 4, the compression strain was denoted as positive while the tensile strain is negative. In the case where individual sensors are installed at the neutral axis of the cross section, the effect of the biaxial bending moment does not need to be taken into consideration; thus, the effect of the bending moment with respect to the y axis $\pm_{My}$ is negligible for the sensors of S1 and S3. Likewise, only the effect of the bending moment with respect to the y axis needs to be taken into consideration for the strain measured at the sensors S2 and S4. Table 1 shows the combined stresses at each sensor location.

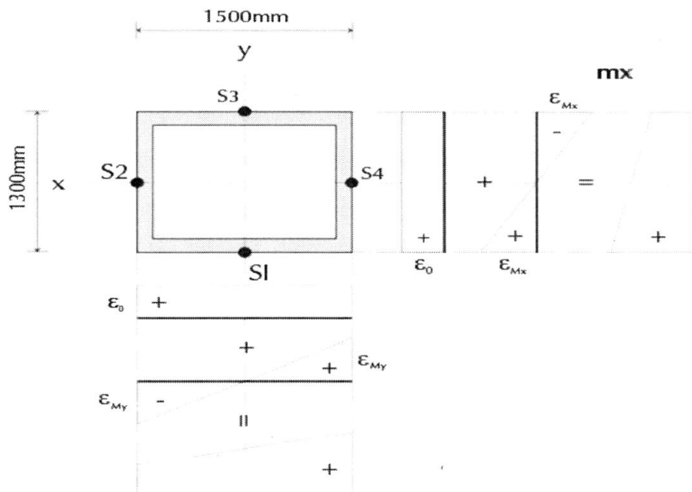

Figure 4: Stress distribution of the mega column subjected simultaneously to an axial force and bending moments.

Table 1: Combined strain at each sensor location

Sensor Location	Axial Strain	Bending Strain	Combined Strain
S1	$\in_o$	$^{\in}M_x$	$\in_1 = \in_0 + ^{\in}M_x$
S2	$\in_o$	$-\in M_y$	$\in_2 = \in_0 - ^{\in}M_y$
S3	$\in_o$	$-\in M_x$	$\in_3 = \in_0 - ^{\in}M_x$
S4	$\in_o$	$\in M_y$	$\in_4 = \in_0 + ^{\in}M_y$

In such cases, three unknown variables are needed to determine the strain at the arbitrary sensors installed on the mega column: $_{o}$, $_{Mx}$, and $_{My}$. Hence, the three unknowns may be determined by randomly choosing three out of the four sensors. With the measurements from the sensors of S1 and S3 or those from the sensors of S2 and S4, the strain due to the axial force, $_0$, can be obtained from Equation (1):

$$\varepsilon_0 = \frac{\varepsilon_1 + \varepsilon_3}{2} = \frac{\varepsilon_2 + \varepsilon_4}{2} \tag{1}$$

If $_0$ is calculated using the strains $_1$ and $_3$ in Equation (1), then M_x can be expressed as in Equation (2), and M_y as in Equation (3) or Equation (4):

$$M_x = {}_1 - \varepsilon_0 = {}_0 - \varepsilon_3 = \frac{\varepsilon_1 - \varepsilon_3}{2} \tag{2}$$

$$\in M_x = \in_0 - \in_2 = \frac{\in_1 + \in_2}{2} - \in_2 \tag{3}$$

$$\in M_y = \in_4 - \in_o = \in_4 - \left(\frac{\in_1 + \in_3}{2}\right) \tag{4}$$

In the same manner, if $_0$ is calculated using the strains $_2$ and $_4$ in Equation (1), then M_x can be expressed as in Equation (5) or Equation (6), and M_y as in Equation (7):

$$\in M_x = \in_1 - \in_o = \in_1 - \frac{\in_2 - \in_4}{2} \tag{5}$$

$$\in M_x = \in_0 - \in_3 = \frac{\in_2 + \in_4}{2} - \in_3 \tag{6}$$

$$\in M_y = \in_0 - \in_2 = \in_4 - \in_0 = \frac{\in_4 - \in_2}{2} \tag{7}$$

(When Equations (1)–(7) are used, even if the measurements have not been normally carried out by one of the four sensors installed to the four sides of the mega column, the non-measured data is recovered based on the strain data normally measured by the other three sensors. For instance, if the strain ${}_1$ was not measured at the sensor S1, the axial strain ${}_0$ can be calculated as the mean of the strains ${}_2$ and ${}_4$ measured at the sensors S2 and S4. With the strains from the sensors of S2 and S4, the remaining unknown's M_x and M_y can be obtained from Equations (6) and (7), respectively. Then, the lost measurement data, ${}_1$, can be recovered according to the first row of Table 1.

APPLICATION TO RECOVERY OF LONGITUDINAL STRAINS IN MEGA COLUMNS

Measurement of Axial Strain during Construction

During the construction of the irregular building structure in Figure 1, axial strain sensing using the wireless VWSG-based strain sensing system in Figure 2 was carried out for one year and five months between 23 July 2010, and 22 February 2012. The plot in Figure 5 shows the measurement values over the total sensing period of the four VWSGs, S1–S4, installed on the mega column. A total of 21,808 measurement values were obtained from the four VWSGs during the sensing period. For the monitoring, the distinctive procedures that took place during the sensing period can be classified into five types of events:

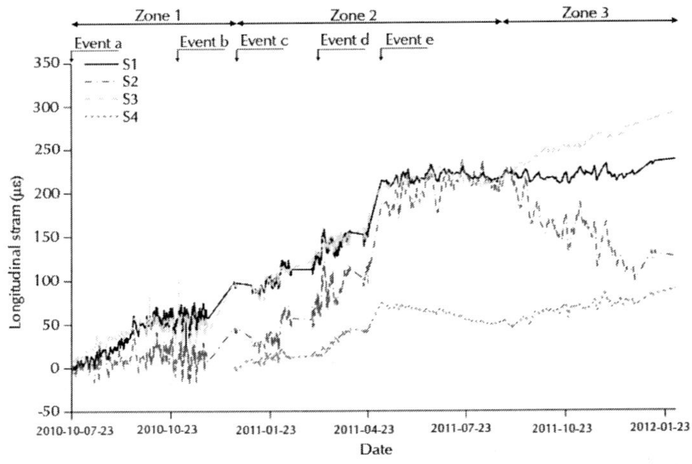

a. Installation of VWSGs (S1, S2, S3)

b. Installation of the edge truss block, 30 October 2010

c. Installation of VWSG S4, 20 December 2010

d. Pouring concrete on the 4th floor cantilever slab, 11 March 2011

e. Removal of supporters (bend in Figure 1) for the edge truss, 25 April 2011

Figure 5: Measurement histories of axial strains in the mega-column during construction.

As shown in Figure 5, the structural response of the mega-column is directly observed through the changes in the measurement values caused by the procedures during the construction period, including the addition of materials and increase of the self-loading, which were reflected by the measurement values and plots obtained from monitoring.

The overall monitoring period was divided into three zones with reference to the major events. Zone 1 is the interval in which, among the four measurement sensors, the measurement sensor S4 was not installed from 23 July 2010 to 20 December 2010. Zone 2 is the interval from 20 December 2010 to 28 August 2011 during which all the measurement sensors (S1~S4) operated normally. Finally, Zone 3 is the interval in which there was an anomaly in the S2 measurement sensor from 28 August 2011 to 22 February 2012.

Recovery of Strain Data

In this study, the measurement data of the S4 sensor, which was not installed in Zone 1, were reconstructed with the data of the other three installed sensors. In Zone 2, the S4 sensor measurement data, which were newly started for measurements, were converted into the strains accumulated from the initial measurements using the S4 data reconstructed in Zone 1. The normally obtained data of the four VWSGs were used to compare the values of one sensor predicted on the basis of the data of three randomly chosen sensors with their actual measurement values. In Zone 3 in which the S2 sensor showed an anomaly, the S2 sensor values were modified using the data of the S1, S3, and S4 sensors.

Zone 1

Only three sensors out of a total of four sensors were installed in Zone 1 during the entire sensing period. Hence, the strain $_4$ at the S4 sensor location, which could not be measured, was estimated. To determine the strain values of the S4 sensor that was not installed, the values of $_0$ and $_{My}$ were required, as shown in the fourth row of Table 1. As in Equation (1), $_0$ is calculated as the mean of $_1$ and $_3$. According to Equation (3), $_{My}$ is determined with the values of $_1$, $_2$, and $_3$. The strain values at the S1, S2, and S3 sensors as well as the $_0$, $_{My}$, and $_4$ values estimated using the strain values at the three sensors are shown in Table 2 at monthly intervals from the first day of the sensing. Figure 6 shows the S1–S3 measurements that were obtained through the constant monitoring and the S4 estimations during the period from 23 July 2010, the day when the monitoring started, to the time when the S4 sensor was installed.

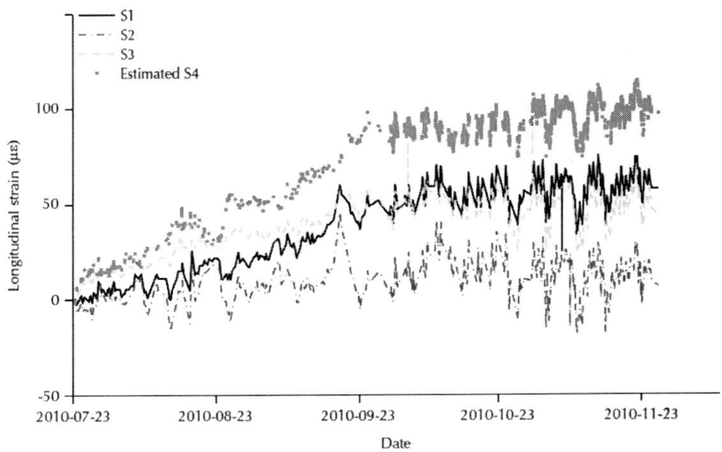

Figure 6: Estimated data for S4 and measurement data for S1, S2, and S3 (Zone 1).

Table 2: Measured strains and estimated strain of $_4$ in Zone 1

Time	Measured Strains($\mu_\in$)			Estimated Strains ($\mu_\in$		Combined Strain ($\mu_\in$)
	$\in_1$	$\in_2$	$\in_3$	$\in_0$	$\in_{M_y}$	$\in_4$
2010-07-23	0.00	0.00	0.00	0.00	0.00	−0.85
2010-07-23	−1.62	−2.36	−1.58	−1.60	0.75	−0.85
2010-08-23	20.84	17.34	27.34	24.09	6.74	30.83
2010-08-23	20.07	−5.48	28.11	24.09	9.10	33.19
2010-09-23	36.88	0.81	40.13	38.51	43.99	82.50
2010-09-23	42.45	34.61	44.15	43.30	42.49	85.79
2010-10-24	69.58	35.38	57.72	63.65	29.04	92.69
2010-10-24	69.58	6.34	57.72	50.08	28.27	91.92
2010-11-26	57.64	6.34	42.53	50.08	43.75	93.83
2010-11-26	57.64	7.88	41.76	49.70	41.82	91.52

Zone 2

Figure 7 shows the measurements of the S4 sensor that was installed on 20 December 2010, about five months after the start of the measurements, and the measurements of the sensors S1–S3 n Zone 2. The period represented by Zone 2 is from 20 December 2010, when sensor S4 was installed, to 28 August 2011, when sensor S2 showed an abnormality. The measurements of S4 were different from those of the S1–S3 sensors that had initially been installed. It is necessary to modify the measurement data of S4 in Zone 2 into the strain accumulated from the initial measurement.

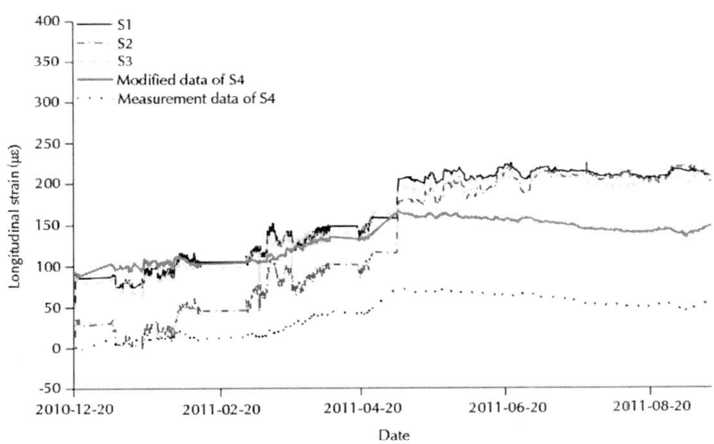

Figure 7: Measured strains from S1, S2, S3, and S4 and modified strain by adding the accumulated strain of 91.52 μ for S4 (Zone 2).

The starting value of S4 in Zone 2, estimated through the S1, S2, and S3 measurements in Zone 2, was compared with the final value of S4 in Zone 1 estimated using the S1, S2, and S3 measurements in Zone 1. The estimated value for S4 value in Zone 1 was 91.52 μ, and that in Zone 2 was 96.31 μ. With the consistency of about 95% in the two values, the measurement data on the day of 20 December 2010 when the S4 sensor was installed were transferred in parallel as much as the S4 value which was 91.52 μ In Figure 7, the S4 measurement values in

Zone 2 were modified by considering the estimation of the S4 sensor at the last period of Zone 1.

The variance accounted for (VAF) in Equation (8) was applied to investigate the accuracy of the estimations and measurements [26, 27]:

$$VAF = \left[1 - \frac{var(y_{mc} - y_{es})}{var(y_{mc})}\right] \times 100 \tag{8}$$

The estimated and measured data are denoted as yes and yme, respectively. "var" denotes variance. Table 3 shows the result of the quantitative evaluation of the estimations and measurements, indicating that the values of S1, S2, and S3 are all consistent by more than 98%.

Table 3: The VAF of each sensor

	S1 (%)	S2 (%)	S3 (%)
VAF of Zone 2	99.36	98.69	99.09

In Figure 8, for example, the estimated strain data based on the S2, S3, and S4 sensors are compared the measured data from S1. It is found that the estimated data matched well with the measured data.

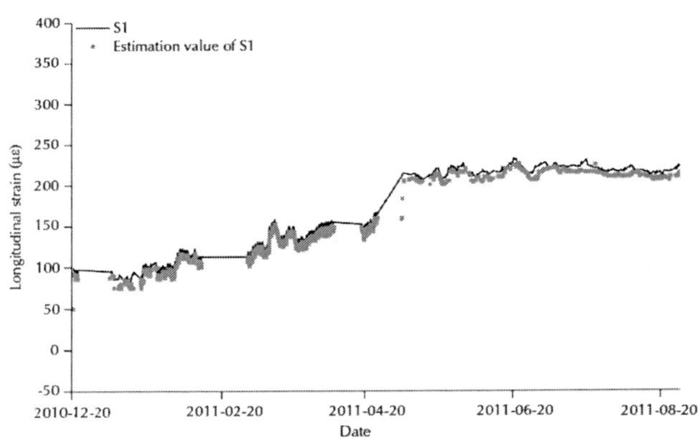

Figure 8: Comparison of estimated and measured strain data for S1 (Zone 2).

Zone 3

In Zone 3, where the anomaly of the S2 sensor was found, increases in the measurements due to the increased load were found in the S1, S3, and S4 sensors. However, the measurements of the S2 sensor were decreasing abnormally in Zone 3 as shown in Figure 9. The reason for the S2 measurement anomaly is not known. Thus, in Figure 9, the S2 value was estimated on the basis of the measurement data of the S1, S3 and S4 sensors that were normally operating so that the estimated S2 value could replace the previous measurement data. Finally, Figure 10 shows the estimation values of the S4 sensor that was not installed in Zone 1, the parallel transfer of the S4 value in Zone 2, and the modified measurement of the abnormal S2 sensor in Zone 3.

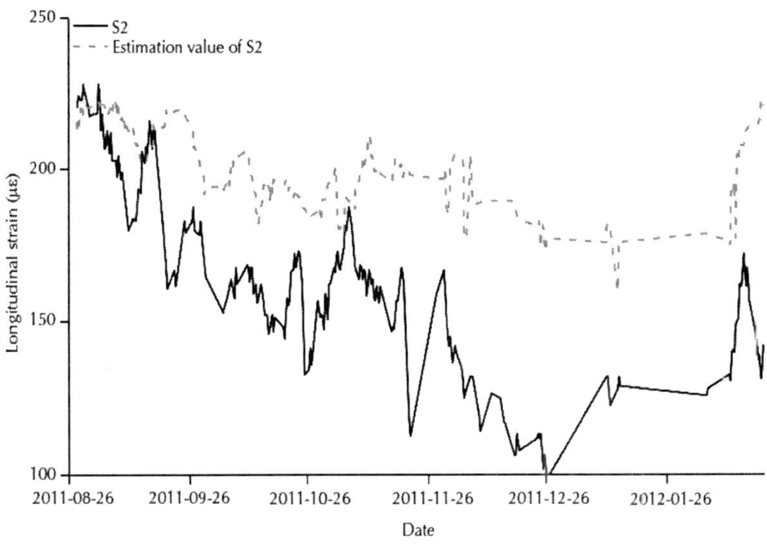

Figure 9: Estimated and measured strain data for S2 (Zone 3).

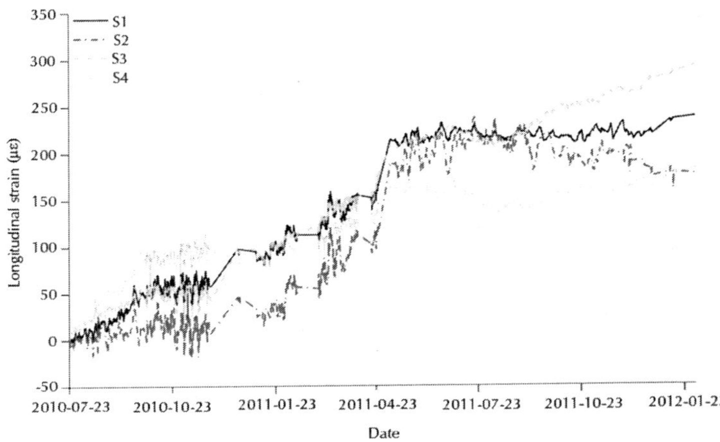

Figure 10: Strain histories for the mega column based on the recovery technique during construction.

CONCLUSIONS

In this study, a strain data recovery technique is proposed for the data losses during structural health monitoring of a mega column in a large-scale irregular building structure. The monitoring was done for one year and five months. During the monitoring, problems such as non-installation of a measurement sensor and abnormal sensor operation occurred. The entire period was divided into three zones: Zone 1 when the S4 sensor had not been installed, Zone 2 when the monitoring was performed normally because the S1–S4 sensors were installed and Zone 3 when the anomaly of the S2 sensor was found. Based on the analytical relations among the measured stains, the measurements of the S4 sensor, which had not been installed in Zone 1, were estimated with a simple estimation method. Additionally, the values measured after the installation of the S4 sensor were transferred in parallel as much as the S4 value measured in Zone 2, and then the structural response of the S1–S4 sensors was monitored. Finally, the measurement of the S2 sensor in which the anomaly was found in Zone 3 was replaced by the estimation based on the S1, S3 and S4 SENSOR measurements. The problems that may occur during long-term wireless sensor monitoring during construction could be resolved through the estimation method. In addition, the safety of structural members during the monitoring may be evaluated by estimating the axial force and moments of the structural members. It is noted that analytical correlation among the strains from strain sensors is required for the application of the data recovery method.

This work was supported by the National Research Foundation of Korea (NRF) grant funded by the Korea government (Ministry of Education, Science and Technology, MEST) (No. 2011-0018360).

REFERENCES

1. Taranath, B.S. Structural Analysis and Design of Tall Buildings: Steel and Composite Construction; CRC Press: Boca Raton, FL USA, 2012; pp. 79–104
2. Park, H.S.; Hong, K.; Seo, J.H. Drift design of steel-frame shear-wall systems for tall buildings. Struct. Des. Tall Spec. Build. 2002, 11, 35–49.
3. Su, J.Z.; Xia, Y.; Chen, L.; Zhao, X.; Zhang, Q.L.; Xu, Y.L.; Chen, A.R. Long-term structural performance monitoring system for the Shanghai Tower. J. Civil. Struct. Health Monit. 2013, 3, 49–61
4. Ji, L.; Zhang, J. The Application of GPRS Remote Wireless Automatic Monitoring System to Health Monitoring in Guangdong Science Center. Proceedings of the Second International Conference on Mechanic Automation and Control Engineering, Hohhot, China, 15–17 July 2011.
5. Park, H.S.; Son, S.; Choi, S.W.; Kim, Y. Wireless laser range finder system for vertical displacement monitoring of mega-trusses during construction. Sensors 2013, 13, 5976–5813
6. Chan, P.K.C.; Jin, W.; Gong, J.M.; Demokan, M.S. Multiplexing of fiber Bragg grating sensors using an FMCW technique. IEEE Photonic. Tech. Lett. 1999, 11, 1470–1472
7. Ansari, F. Practical implementation of optical fiber sensors in civil structural health monitoring. J. Intell. Mater. Syst. Struct. 2007, 18, 879–889

8. Park, H.S.; Lee, H.M.; Adeli, H.; Lee, I. A new approach for health monitoring of structures: Terrestrial laser scanning. Comput. Aided Civil. Infrastr. Eng. 2007, 22, 19–30
9. Choi, S.W.; Lee, J.; Kim, J.M.; Park, H.S. Design and application of a field sensing system for ground anchors in slopes. Sensors 2013, 13, 3739–3752
10. Park, H.S.; Shin, Y.; Choi, S.W.; Kim, Y. An integrative structural health monitoring system for the local/global responses of a large-scale irregular building under construction. Sensors 2013, 13, 9085–9103
11. Alamusi, H.N.; Fukunaga, H.; Atobe, S.; Liu, Y.; Li, J. Piezoresistive strain sensors made from carbon nanotubes based polymer nanocomposites. Sensors 2011, 11, 10691–10723
12. Hu, B.; Hu, N.; Cai, Y.; Yuan, W.; Yan, C.; Cai, Y.; Furukawa, M.; Matsushita, M. Performance characterization of VGCF/epoxy nanocomposite sensors under multiple load cycles and in static structural health monitoring. Smart. Mater. Struct. 2013, 22
13. Hu, N.; Karube, Y.; Arai, M.; Watanabe, T.; Yan, C.; Li, Y.; Liu, Y.; Fukunaga, H. Investigation on sensitivity of a polymer/carbon nanotube composite strain sensor. Carbon 2000, 48, 680–687
14. Nair, R.V.; Vijaya, R. Photonic crystal sensors: An overview. Prog. Quant. Electron. 2010, 34, 89–134
15. Lee, H.M.; Kim, J.M.; Sho, K.; Park, H.S. A wireless vibrating wire sensor node for continuous structural health monitoring. Smart. Mater. Struct. 2010, 19
16. Bao, Y.; Li, H.; Sun, X.; Yu, Y.; Ou, J. Compressive sampling-based data loss recovery for wireless sensor networks used in civil structural health monitoring. Struct. Health Monit. 2013, 12, 78–95
17. Lee, J.-H.; Jung, I.-B. Adaptive-compression based congestion control technique for wireless sensor networks. Sensors 2010, 10, 2919–2945
18. Kim, S.; Brendle, C.; Lee, H.-Y.; Walter, M.; Gloeggler, S.; Krueger, S.; Leonhardt, S. Evaluation of a 433 MHz band body sensor network for biomedical applications. Sensors 2013, 13, 898–917

19. Qiu, X.; Liu, H.; Li, D.; Yick, J.; Ghosal, D.; Mukherjee, B. Efficient aggregation of multiple classes of information in wireless sensor networks. Sensors 2009, 9, 8083–8108
20. Lee, J.-H.; Jung, I.-B. Speedy routing recovery protocol for large failure tolerance in wireless sensor networks. Sensors 2010, 10. 3389–3410
21. Kang, I.; Schulz, M.J.; Kim, J.H.; Shanov, V.; Shi, D. A carbon nanotube strain sensor for structural health monitoring. Smart. Mater. Struct. 2006, 15, 737–748. [Google Scholar]
22. Efendioglu, H.S.; Yildirim, T.; Fidanboylu, K. Prediction of force measurements of a microbend sensor based on an artificial neural network. Sensors 2009, 9, 7167–7176. [Google Scholar]
23. Baladrón, C.; Aguiar, J.M.; Calavia, L.; Carro, B.; Sánchez-Esguevillas, A.; Hernández, L. Performance study of the application of artificial neural networks to the completion and prediction of data retrieved by underwater sensors. Sensors 2012, 12, 1468–1481. [Google Scholar]
24. Geokon. Available online: http://www.geokon.com/strain-gages/ (accessed on 27 May 2013).
25. Texas Instruments. Single Chip Very Low Power RF Transceiver with 8051-Compatible Microcontroller, Available online: http://www.ti.com/product/cc1010 (accessed on 28 May 2013).
26. Mittal, H.; Bhatia, P. Optimization criteria for effort estimation using fuzzy technique. CLEI Electron. J. 2007, 10, 1–11
27. Mitchell, R.; Kim, Y.; El-Korchi, T. System identification of smart structures using a wavelet neuro-fuzzy model. Smart Mater. Struct. 2012, 21

Chapter 2

Applications of Sma Bundles in Practical Concrete Structures

Lei Li[1, 2], Qingbin Li[1] and Fan Zhang[3]

[1] State Key Laboratory of Hydroscience and Engineering, Tsinghua Univ., Beijing, P. R. China

[2] China Three Gorges Corporation, Beijing, P. R. China

[3] Dept. of Transportation of Henan Province, Zhengzhou, P. R. China

INTRODUCTION

In the recent twenty years, SMA composites were studied, and the matrix was almost polymer or fiber reinforced resin. During the recent ten years, researches on applying SMA in concrete structures were carried out, but yet limited in laboratory.

In this chapter, the constitutive characteristics of NiTi SMA and its application in practical concrete structures were investigated. Firstly, the transformation characters of the NiTi SMA adopted in this study were obtained through a differential scanning calorimeter (DSC) analysis technology, and then the properties of NiTi SMA during uncompleted transformation process and the effects of plastic deformation on the transformation were studied. The uniaxial tension, SME, and the constrained recovery process of NiTi SMA were examined through an improved 10KN universal material testing machine, effects of the constraint conditions, the maximum tension deformation and the initial phase of SMA on the constrained recovery process were compared and investigated.

To simulate the characteristics of NiTi SMA more effectively, a new constitutive model derived from the internal variable approach was constructed based on the DSC and the uniaxial tension experimental results, a new simple kinetics equation was presented, and the plastic deformation was considered in the constitutive equation.

As the application of SMA in practical concrete structures, a new concept for designing a smart concrete bridge in a freeway was presented against the overload problem in the transport field, and SMA was firstly applied in controlling the deformation of a concrete beam bridge, performances of the concrete beams and the concrete bridge with embedded SMA were examined through a sophisticated test program.

One-Dimensional Constitutive Model of SMA with an Empirical Kinetics Equation

The characteristics of NiTi SMA were examined through sophisticated testing program, and the constitutive model of NiTi SMA with an empirical kinetics equation was investigated. Firstly, the transformation characters of the NiTi SMA were obtained through a differential scanning calorimetry (DSC) analysis technology, and the properties during incomplete and discontinuous transformation process and the effects of plastic deformation on the transformation were studied. The uniaxial tension, SME and constrained recovery process of NiTi SMA were examined through an improved 10KN universal material testing machine. Experimental results indicated that the phase transformation characters and the mechanical properties could be affected by the loading process considerably, and the plastic deformation should be taken into account. To simulate the characteristics of NiTi SMA more effectively, a one-dimensional constitutive model derived from the internal variable approach with the consideration of the plastic deformation was constructed based on the DSC and the uniaxial tension experimental results, in which a new simple empirical kinetics equation was presented, and the transformation temperature parameters were redefined according to the DSC experimental evidence. Comparison between the numerical and experimental results indicated that this constitutive model could simulate the phase transformation characters, the uniaxial tension, SME and the constrained recovery behavior of NiTi SMA well.

BACKGROUND

Since the Shape Memory Effect (SME for short) was observed in Cu-Zn alloy (Greninger & Mooradian, 1938), and in NiTi alloy (Buehler, Gilfrich, & Wiley, 1963), SMA (especially NiTi alloy) has been widely used as intelligent material for its particular characteristics, such as large load capacity, high recovery strain up to 8%, excellent fatigue performance, variable elastic modulus with phase transformation, and especially the two main interesting properties, SME and Pseudo-elasticity (PE) due to the diffusionless martensitic transformation, as discussed elsewhere (Delaey et al., 1974;Krishnan et al., 1974; Warlimont et al., 1974; Funakubo, 1987; Otsuka & Wayman, 1998).

To simulate these specific properties, many constitutive laws have been proposed, such as the phenomenological models (Tanaka & Nagaki, 1982; Tanaka, 1986; Liang, 1990; Brinson, 1993; Zhu et al., 2002), micromechanics models (Sun & Hwang, 1993a, 1993b), 3D model for polycrystalline SMA based on microplane theory (Brocca, Brinson, & Bazant, 2002), etc. Among them, phenomenological models based on the internal variable method were the most popular used in the practical engineering. The mechanical constitutive equation was derived from the principles of thermodynamics, martensite fraction as an internal variable was used to represent the stage of the transformation, and an empirical kinetics equation was proposed to describe this transformation governed by temperature and stress. The major difference of these models was its specific kinetics equation.

The first phenomenological model was derived by Tanaka and Nagaki (1982) from the first and second laws of thermodynamics and can be written as

$$\sigma = \rho_0 \frac{\partial \phi}{\partial \varepsilon} = \sigma(\varepsilon, T, \xi) \tag{1}$$

Where σ, ρ_0, Φ, ε, T and ξ represent the second Piola-Kirchhoff stress, the density, Helmholtz free energy, Green strain, temperature and the martensite fraction, respectively. Equation (1) can be written by differential calculus as

$$\dot{\sigma} = \frac{\partial \sigma}{\partial \varepsilon}\dot{\varepsilon} + \frac{\partial \sigma}{\partial T}\dot{T} + \frac{\partial \sigma}{\partial \xi}\dot{\xi} = D\dot{\varepsilon} + \Theta\dot{T} + \Omega\dot{\xi} \tag{2}$$

Where D= $\rho_0 \frac{\partial^2 \Phi}{\partial \varepsilon^2}$ is the modulus of the SMA materials, Θ= $\rho_o \frac{\rho^2 \Phi}{\partial \varepsilon \partial T}$ is related to the thermal coefficient of expansion, and Ω= $\rho_o \frac{\rho^2 \Phi}{\partial \varepsilon \partial \xi}$ can be regarded as the "transformation tensor". The relationship between the martensite fraction, the temperature and the stress is expressed by an exponential kinetics equation as

$$\xi_{M \to A} = \exp\left[a_A (T - A_s) + b_A \sigma\right] \tag{3}$$

for the transformation from martensite to austenite, and

$$\xi_{A \to M} = 1 - \exp\left[a_M (T - M_s) + b_M \sigma\right] \tag{4}$$

for the transformation from austenite to martensite, where *aA*, *aM*, *bA*, *bM* are the material constants related to the transformation temperature, *As* and *Ms* are the start temperatures of austenite transformation and the martensite transformation, respectively. Liang (1990) simplified the constitutive model based on Equation (2), material functions were assumed to be constants, and the constitutive relation can be written as

$$\sigma - \sigma_0 = D(\varepsilon - \varepsilon_0) + \Omega(\xi - \xi_0) + \Theta(T - T_0) \tag{5}$$

Where the subscript "0" indicates the initial conditions of the materials. Martensite fraction, as a function of stress and temperature during transformation, is represented by an empirically based cosine models as

$$\xi_{M\to A} = \frac{\xi_0}{2}\cos\left[a_A(T - A_s) + b_A\sigma\right] + \frac{\xi_0}{2} \tag{6}$$

for the transformation from martensite to austenite, and

$$\xi_{A\to M} = \frac{1-\xi_0}{2}\cos\left[a_M(T - M_f) + b_M\sigma\right] + \frac{1+\xi_0}{2} \tag{7}$$

for the transformation from austenite to martensite, where ξ_0 is the initial martensite fraction, a_A, a_M, b_A, b_M are the material constants, *As* is the start temperature of austenite transformation, *Mf* is the finish temperature of martensite transformation. Brinson (1993) redefined the martensite fraction based on the micromechanics of SMA material as

$$\xi = \xi_S + \xi_T \tag{8}$$

Where ξ_s represents the fraction of the stress-induced martensite (or single-crystal martensite) and ξ_T denotes the fraction of the temperature-induced martensite with multiple variants. Young's modulus was assumed to be non-constant from the experimental evidence as

$$D(\varepsilon, \xi, T) = D(\xi) = D_A + \xi(D_M - D_A) \tag{9}$$

Where *DA* and *DM* are the Young's moduli of austenite and martensite. The transformation tensor was also redefined as

$$\Omega(\xi) = \Omega(\xi_0) + (\xi_0)\Omega'(\xi_0) \tag{10}$$

Because of these improvements, phenomenological models based on the internal variable method could simulate the characteristics of SMA well, such as SME and PE, and have been used in some commercial program for its simple form. Whereas, DSC experiment indicated that there were no apparent transformation start and finish temperatures. And additionally, for the SMA materials used as actuators in practical engineering, not only the stress but also the thermo load determined the transformation process simultaneously, especially the latter. In this section, constitutive model was presented to simulate the characteristics of SMA materials for its practical utilization based on the former effort and the thermal and mechanical experiments. Transformation temperature parameters were redefined, a new simple empirical kinetics equation was presented, and a mechanical equation was developed to describe its mechanical character more directly.

EXPERIMENTS AND RESULTS

Thermodynamics Experiment

Complete transformation

The Ti-49.5 wt % Ni in this research (binary, straight annealed) was one-way SMA. Its transformation property can be achieved through a DSC apparatus (type: Mettler Toledo DSC821 e), as shown in Fig. 1, the four transformation temperatures can be determined as: martensite transformation start and finish temperatures denoted by M_s and M_f are 33.8°C and 23.3 °C, austenite transformation start and finish ones (*As* and A_f) are 41.9 °C and 59.6 °C, respectively. It should be noticed that these transformation start and finish temperatures were defined through the DSC diagram artificially, actually the transformation could take place slowly earlier than the transformation start temperature and remain unfinished after the transformation finish temperature, as shown in the experimental curve, thus the recovery forces induced by the transformation might increase or decrease beyond the transformation start and finish temperatures. Accordingly, a new series of parameters were adopted to describe this transformation character more realistically, where T_A and T_M denotes the temperatures corresponding to the peak points of the austenite and martensite transformations, A and M indicate the widths of the transformation peaks, A≈(Af−As)/2e and M≈(Ms−Mf)/2e , and e is the natural constant and equals to 2.71828. Subtracting the exothermic or the endothermic part of non-phase transformation process from the DSC curve (as seen in Fig. 1), normalizing the heat absorbing and releasing capacities, and taking the absolute value, martensite quality fraction ratio versus temperature can be achieved, as shown in Fig. 2. Calculating the integral of the martensite quality fraction ratio, the relationship of martensite quality fraction versus temperature during the martensite and austenite transformation process can be obtained, as shown in Fig. 3.

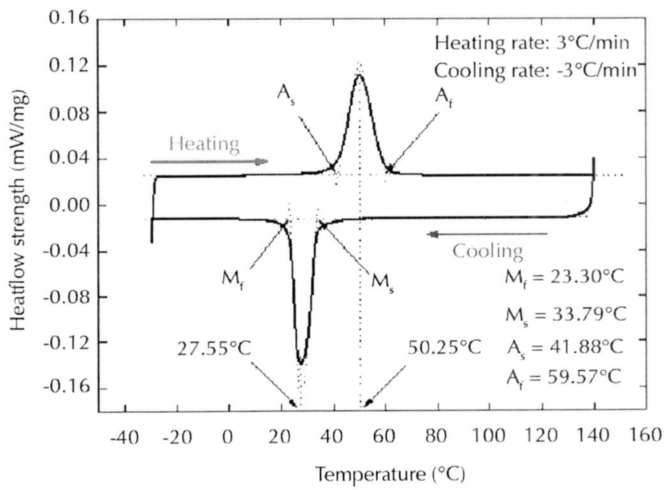

Figure 1: DSC diagram of SMA.

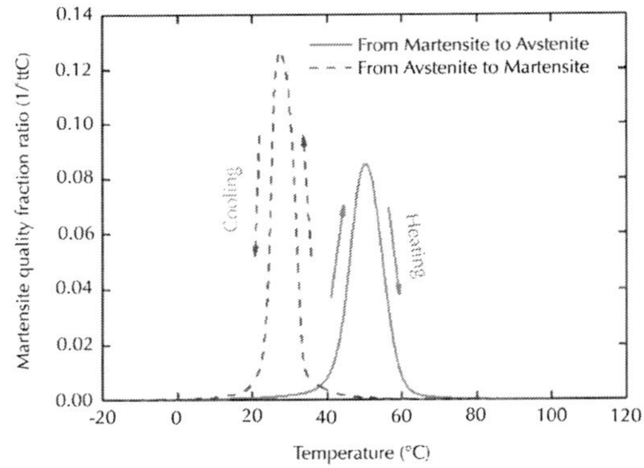

Figure 2: Variation rate of martensite quality fraction versus temperature.

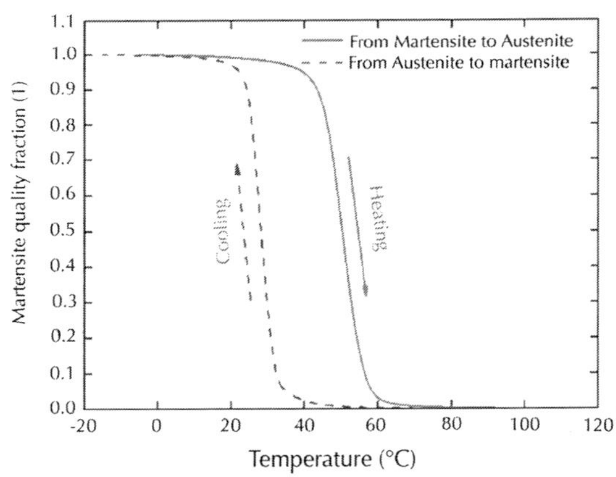

Figure 3: Variation of martensite quality fraction versus temperature.

Incomplete and Discontinuous Transformation

If the initial state is not the complete martensite or austenite, the phase transformation will be different. At the temperature of 140°C, the initial state of SMA is pure austenite, cooling will induce the transformation from austenite to martensite, while at half of this transformation, heating will lead to the reverse transformation from the new generated martensite to austenite, the start and finish temperature and the shape of the peak are similar to the ones of the complete austenite transformation, but the peak value is smaller than the complete transformation one, as shown in Fig. 4. Similarly, at the temperature of -30 °C, the initial state is pure martensite, heating will induce the transformation from martensite to austenite, while at half of the transformation, cooling will lead to the transformation from the new generated austenite to martensite, the start and finish temperatures and the shape of the peak are similar to those of the complete martensite transformation, but the peak value becomes smaller than the complete transformation one, as shown in Fig. 5. These phenomena indicate that the martensite transformation intensity is related to the initial austenite quality fraction and the austenite transformation intensity is related to the initial martensite quality fraction. Accordingly, it is assumed that the martensite transformation intensity is proportional to the initial austenite fraction, and the austenite transformation intensity is proportional to the initial martensite fraction.

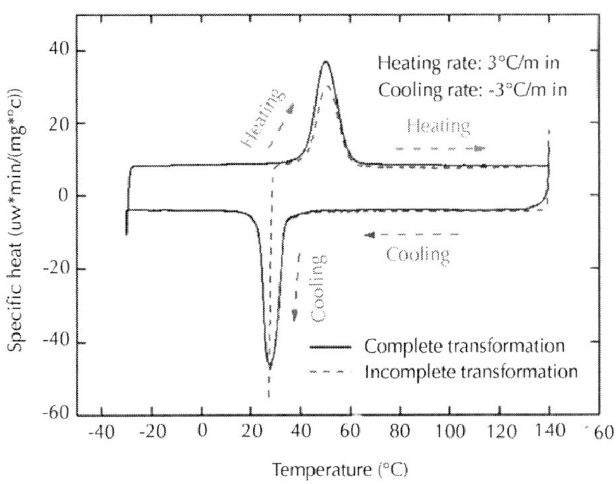

Figure 4: Incomplete transformation from martensite to austenite.

During the transformation from austenite to martensite, i.e., the cooling process, if the small heating does not induce the transformation from martensite to austenite, the martensite transformation will continue during the subsequent cooling process, as shown in Fig. 6.

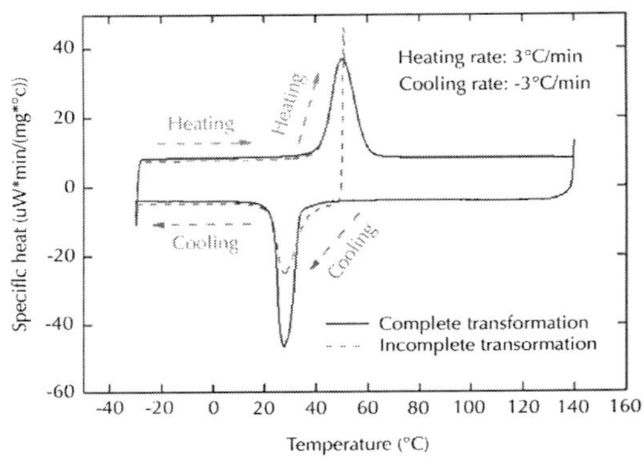

Figure 5: Incomplete transformation from austenite to martensite.

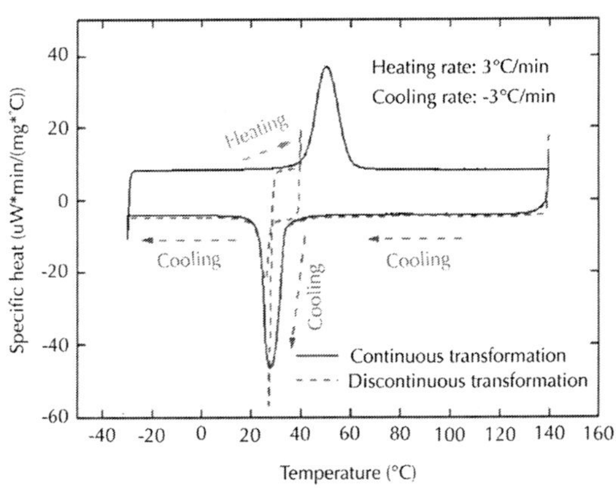

Figure 6: Discontinuous transformation from austenite to martensite.

Similarly, during the transformation from martensite to austenite, i.e., the heating process, if the small cooling does not induce the transformation from austenite to martensite, the austenite transformation will continue during the subsequent heating process, as shown in Fig. 7.

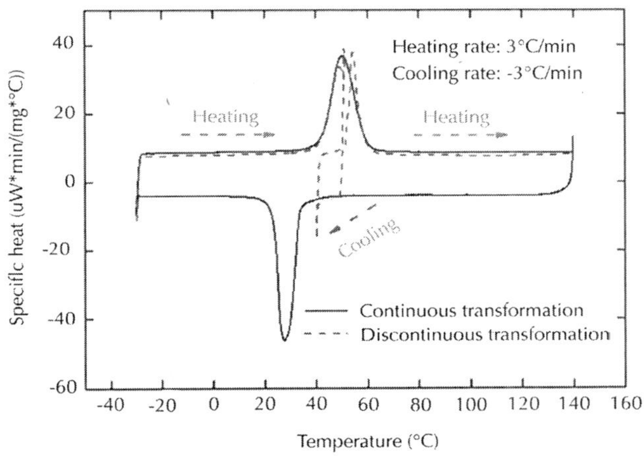

Figure 7: Discontinuous transformation from martensite to austenite.

Influence of Plastic Deformation

Plastic deformation can also change the transformation properties of SMA materials. Stretch SMA to its strain limit, great plastic deformation will be induced, and the DSC experimental result can be achieved as shown in Fig. 8. The temperatures of the austenite transformation and martensite transformations become higher, the width of the peak become wider, and the peak value become smaller than these before tension. This phenomenon comes from the variation of the crystal structure and the internal stress induced by the plastic deformation. Actually, the transformation properties of the SMA crystal cell did not change, just the micro structure of SMA material varied.

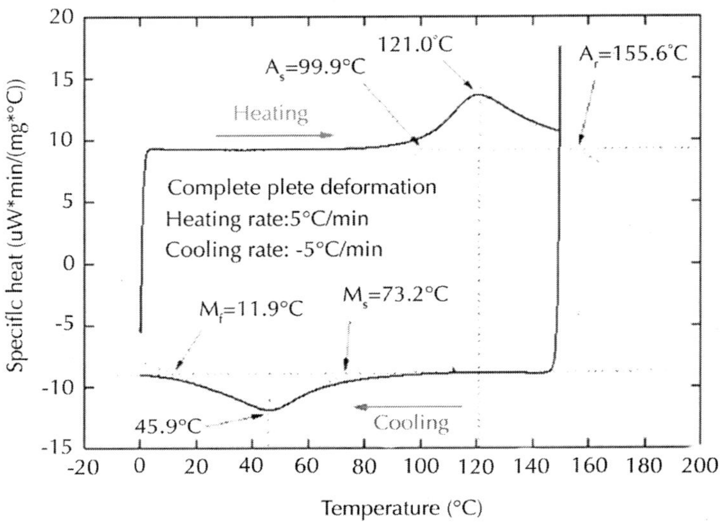

Figure 8: DSC analysis of plastic deformed SMA.

Mechanics Experiment

Mechanical experiment was carried out to determine the material characteristics and parameters through a WDW-10 universal material test machine. The experimental equipment is shown in Fig. 9. Pt100 platinum electric resistance temperature sensors were plastered on the SMA wire with silica gel to monitor its temperature, strain gauge was used to measure its strain. Load applied on the SMA wire, environment temperature, current intensity, and the voltage of SMA wire between the two points of the strain gauge were also monitored at the same time. All these signals were acquired through a 16-channel dynamic data acquisition system.

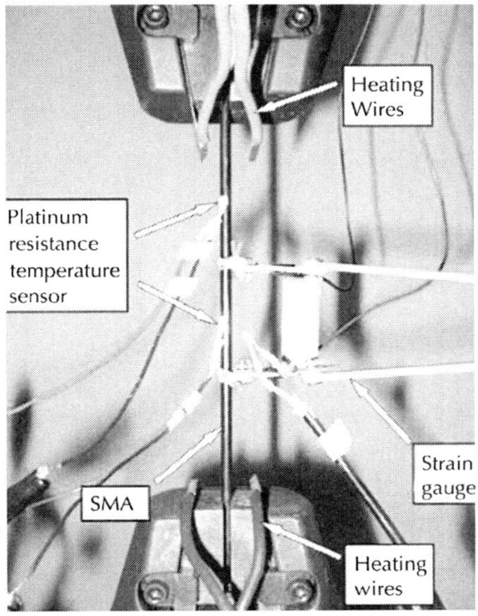

Figure 9: Uniaxial tension test equipment.

Uniaxial Tension Test

The first test was made with constant ambient temperature. SMA wire samples were heated by boiled water firstly, and then cooled by ice water, thus the initial state of the samples was pure twinned martensite. Experimental results are shown in Fig. 10. There are two horizontal "plastic" stages during the tension process of 1 mm diameter SMA wire. The first stage was derived from the detwinning process of the twinned martensite, i.e. from twinned martensite to single martensite, which can be named "pseudo-plasticity" that could be recovered by heating, and the second stage was induced by plastic deformation, which was the "real plasticity". The second uniaxial tension test was carried out with different SMA temperatures. Similarly, SMA wire samples were also pretreated to obtain different initial state. For example, if the sample was heated by boiled water, then cooled by ice water, and then heated to the test temperature of 44.5 °C by electric current, the initial state will be twinned martensite. However, if the sample was heated by boiled water, and then cooled to the test temperature of 44.5 °C, the initial state will be austenite. During the experimental process, testing temperatures were kept constant by adjusting the current intensity.

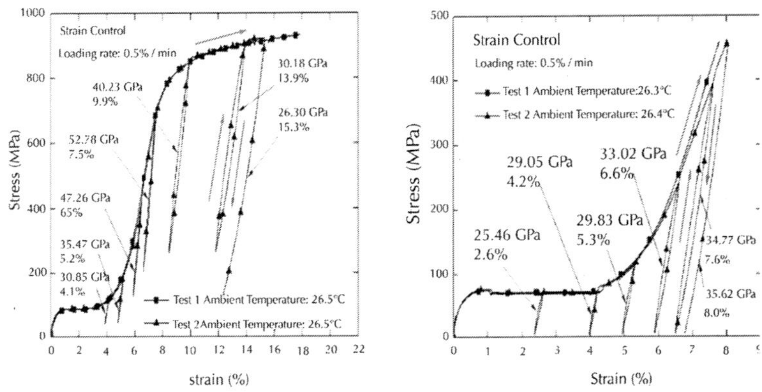

(a). SMA diameter: 1mm (b). SMA diameter: 5mm

Figure 10: Uniaxial tension test.

Elastic moduli of austenite, twinned martensite and detwinned martensite can be determined as 85.5 GPa, 28.24 GPa and 25.75 GPa, respectively based on the experimental result, as shown in Fig. 11, where "M" means the initial state of the sample as twinned martensite, and "A" as austenite. At different ambient temperatures, the initial elastic moduli of austenite are the same. In the ambient temperature of 44.5°C, no matter what the initial state is, the critical stress that leads to the stress-induced martensite transformation or the detwinning process of twinned martensite was almost equal. During the heating recovery process under stress free conditions, the critical temperatures of the thermo-induced austenite transformation were almost the same. Otherwise, the unrecoverable stain of the specimen stretched from martensite initial state was a little bit bigger than that from austenite, as shown in Fig. 12. For the sample starting from complete austenite, the critical stress inducing martensite transformation increased with the temperature. The statistic result is shown in Fig. 13.

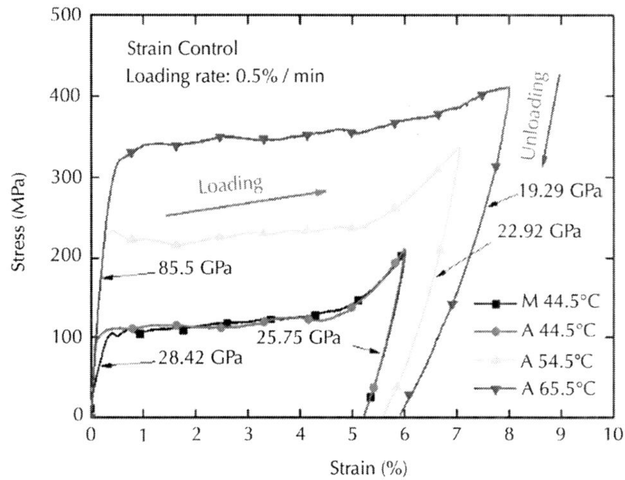

Figure 11: Uniaxial tension processes under different ambient temperatures.

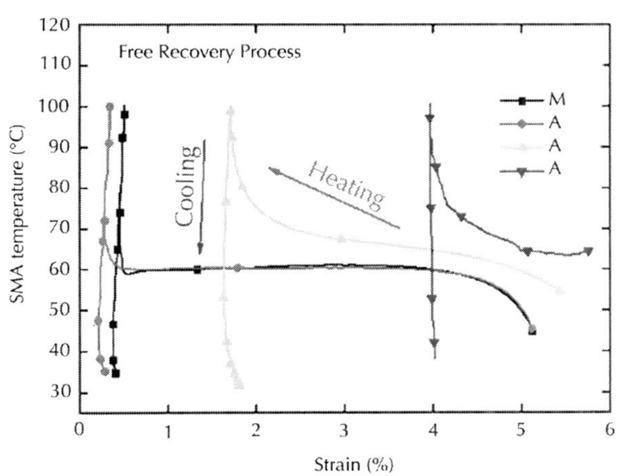

Figure 12: Free recovery processes of SMA afer loading with different temperatures.

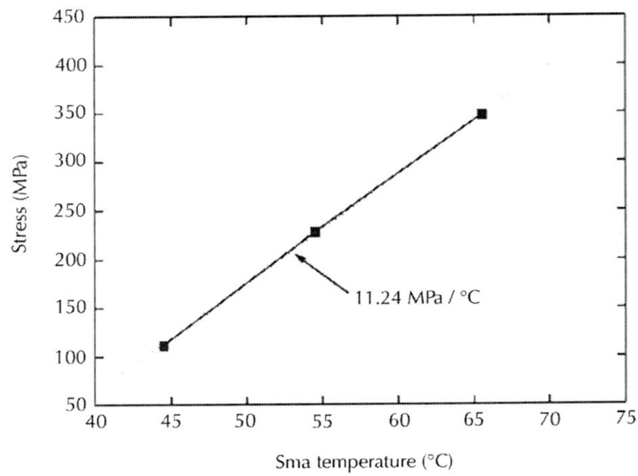

Figure 13: Variation of critical stress with stress induced martensite transformation temperature.

SME test SMA samples with 5mm in diameter were firstly stretched to 4.2%, 5.3%, 6.6%, 7.6%, respectively, and then heated in strain free conditions, this loading cycle was repeated for three times. For the sample with the loading strain of 5.3%, variation of stress with strain during the uniaxial tension process is shown in Fig. 14, and the variation of strain with temperature during the free heating process is shown in Fig. 15. Stress-strain curve of SMA with other maximum loading strains are similar to Fig 14 and Fig. 15. Experimental results indicated that the plastic deformation increased with the loading times, and during the free recovery process, critical temperature inducing the reverse martensite transformation decreased with loading times, and the recoverable strain became smaller and smaller. Comparing the samples with different maximum tension strain, variations of stress with strain and that of strain with temperature during the second loading times and free recovery process are shown in Figs16 and 17, respectively. Test results during the first and the third loading cycles are similar to each other.

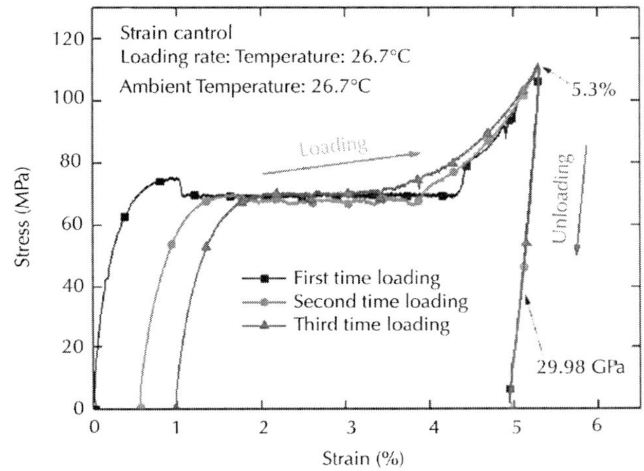

Figure 14: Stress-strain curve of SMA in different loading times

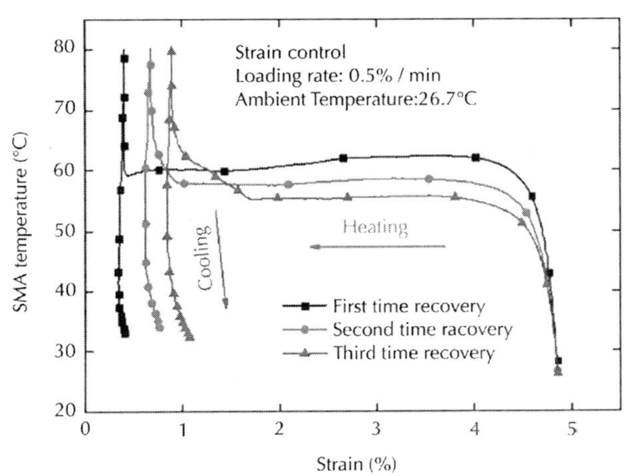

Figure 15: Temperature-strain curve of SMA in different loading times.

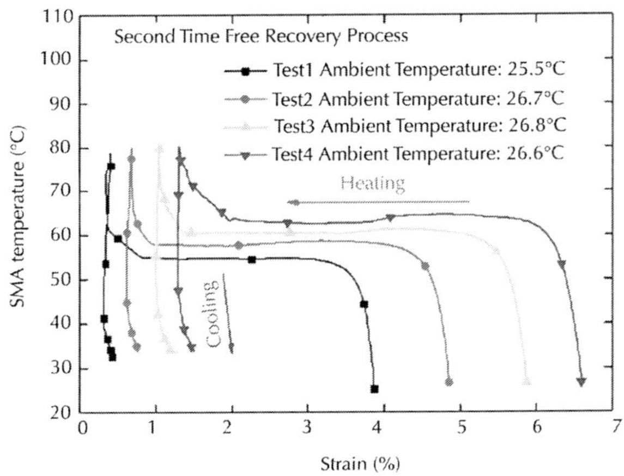

Figure 16: Stress-strain curve of SMA in different maximum loading deformations.

Additionally, statistical results indicated that the critical temperature inducing the reverse martensite transformation increased with the maximum deformation, decreased with the loading/unloading times, as shown in Fig. 18. At the same time, the plastic deformation increased with the maximum deformation and the loading/unloading times, as shown in Fig. 19. According to the experimental curve, the plastic deformation increased with the initial tension strain during the loading and unloading process, and the critical temperature inducing the austenite transformation increased with the initial tension strain, also increased with the plastic deformation.

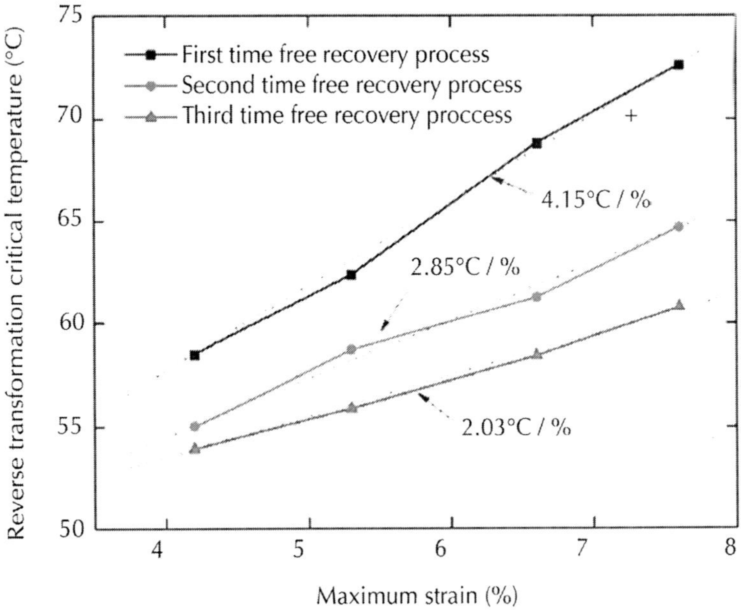

Figure 17: Temperature-strain curve of SMA in different maximum loading deformations.

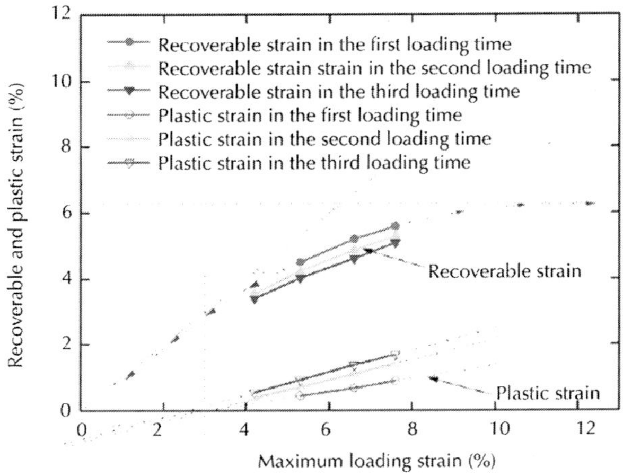

Figure 18: Variation property of transformation critical temperature with maximum loading strain.

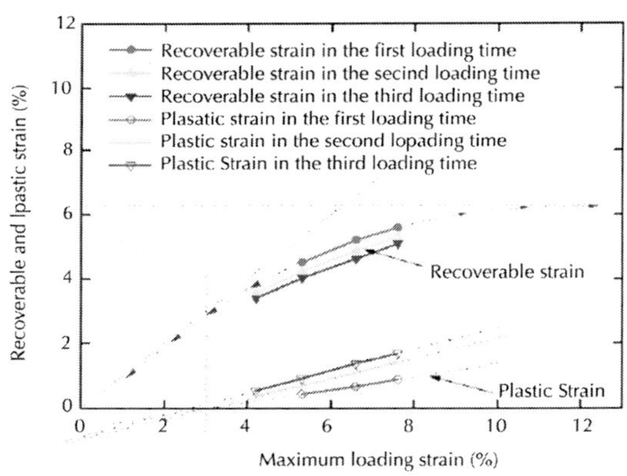

Figure 19: Variation property of recoverable and plastic strain with maximum loading strain.

Recovery Test under Constrained Condition

The recovery test under constrained condition contained five loading paths: (1) strain-control loading to 4.2% under normal temperature; (2) heating to 90 °°C and then cooling to 30 °°C under elastic constrained condition in the first time; (3) heating to 90 °C and then cooling to 30°C under elastic constrained condition in the second, third and fourth times; (4) unloading to free condition; (5) heating to 100 °C and then cooling to 30 °C under free condition. In the initial state of twinned martensite, the elastic constrained recovery test of SMA is shown in Fig. 20. Test results under different constrained conditions, initial state and maximum strain are similar in form.

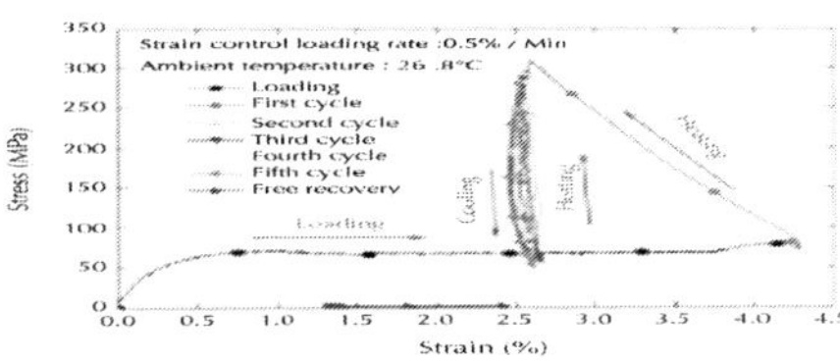

(a). Stress-strain curve

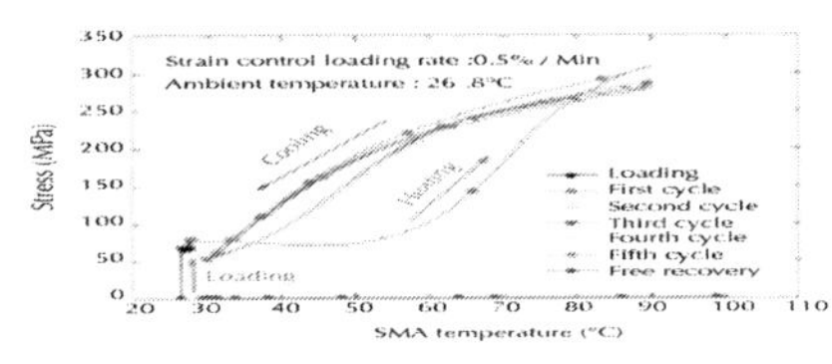

(b).Stress-temperature curve

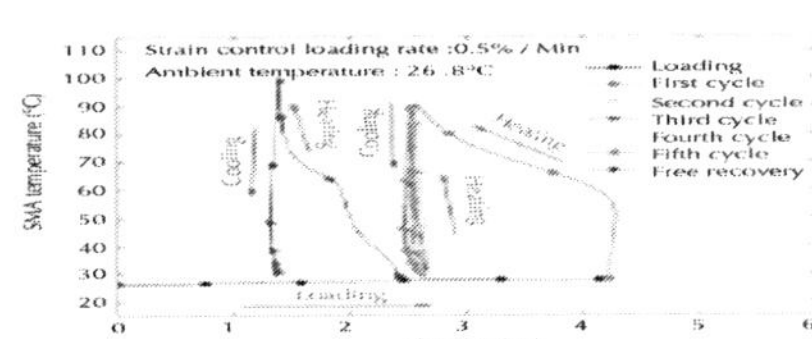

(c). Temperature-strain curve

Figure 20: Constrained recovery test.

Because constraint elastic modulus during the first heating process was smaller than that of the subsequent second to fifth heating/cooling process, relationships between stress and strain, stress and temperature, and temperature and strain, were different correspondingly. In the first heating process, the stress increased gradually, the strain decreased and some deformation of SMA restored. In the following heating and cooling process, the strain almost remained, the stress increased with temperature rising and decreased with temperature dropping, forming a hysteretic circle, and the variation rate of stress with temperature changed in the range of 2.4 MPa ~ 8 MPa / °C. During the heating process under stress free state, the strain recovery course of SMA with temperature showed two obvious stages, as shown in Fig. 20 (c).

Constitutive Model

Physical Equation

According to the macro-phenomenon, deformation of SMA during thermo and stress loading process can be divided into four components: (1) macro recoverable deformation caused by martensite transformation, mainly determined by single crystal martensite percentage; (2) elastic deformation; (3) Thermal expansion deformation; (4) Plastic deformation. Thus, the macro physical equation of SMA in strain form during the temperature and stress loading process can be expressed as

$$= \varepsilon R + \varepsilon E + \varepsilon T + \varepsilon P$$

$$_{res}.\xi_{SM} + \frac{\sigma}{(\xi_{SM}.E_{SM} + \xi_{TM} + \xi_A.E_A)} + (\xi_{SM}.\eta_{SM} + \xi_{TM}.\eta_{TM} + \xi_A.\eta_A).(T - T_0) + \tag{11}$$

where ε is the dimensionless macro total strain, positive for stretch; ε^R , ε^E , ε^T , ε^P denote the dimensionless recoverable transformation, elastic, thermal expansion and plastic strains, respectively; $_{res}$ is the maximum recoverable strain; σ is the macro stress, positive for stretch (unit: MPa). E_{SM} , E_{TM} , E_A denote the elastic moduli of single crystal martensite, twinned martensite and austenite, respectively (units: GPa). η_{SM} , η_{TM} , η_A denote the thermal expansion coefficients of single crystal martensite, twinned martensite and austenite, respectively (units: 1/OC). ξ_{SM} , ξ_{TM} , ξ_A denote the quality fractions of single crystal martensite, twinned martensite and the austenite, respectively, with $\xi_{SM}+\xi_{TM}+\xi_{A}$ equals to 1. T is the temperature of SMA (unit: °C). T_0 is the initial temperature of SMA (unit:°C).

Kinetics Equation

DSC experiments showed that, there were no obvious transformation start and finish temperatures during the transformation process. In this section, the single transformation characteristic temperatures, as well as the parameters that reflect the width of the endothermic and exothermic peaks, are adopted to simulate the phase transformation of SMA, and the corresponding kinetics equation is established at the same time. During the transformation process, there are three crystal structures in SMA material, twinned martensite, single martensite, and austenite, respectively, thus there are six phase transformation processes theoretically. In this section, each phase transition process is analyzed based on the DSC experiments, and the corresponding transformation principles and conditions are presented.

Transformation from Twinned Martensite to Austenite Transformation Condition

Assuming in one-dimensional terms, only heating can lead to this phase transition from twinned martensite to austenite. Transformation principle: For the transformation from twinned martensite to austenite, if the reverse transformation from austenite to twinned martensite did not occur during the process, a small cooling will not affect its subsequent transformation process, as shown in Fig. 7. However, if the reverse transformation process happened and induced a lot of austenitic to twinned martensite, then the extent of the following transformation from twinned martensite to austenite required reduction in accordance with the quality fraction of the current twinned martensite, as shown in Fig. 4. Thus, the kinetics equation from twinned martensite to austenite can be acquired as

$$\xi_{TM} = \frac{\xi TM0}{1+e\dfrac{T-T_A}{A}} \tag{12}$$

where TM0 is the initial quality fraction of twinned martensite; TA is the characteristic temperature for austenite transformation, and numerically equivalent to the temperature where the complete austenite transformation goes into the half, and is equivalent to the peak temperature of the DSC endothermic curve approximately; A is the transformation parameter reflecting the width of the endothermic peak, and numerically equivalent $\frac{A_f - A_s}{2e}$, here A_s A_f denote the start and finish temperatures of austenite transformation defined by the traditional method. e is the natural constant and equals to 2.71828.

Transformation from Austenite to Twinned Martensite

Transformation condition: Assuming in one-dimensional terms, only cooling can lead to this phase transition from austenite to twinned martensite. Transformation principle: For the transformation from austenite to twinned martensite, if the reverse transformation from twinned martensite to austenite did not occur during the process, a small heating will not affect its subsequent transformation process, as shown in Fig. 6. However, if the reverse transformation process happened and induced a lot of twinned martensite to austenitic, the extent of the following transformation from austenite to twinned martensite required reduction in accordance with the quality fraction of the current austenite, as shown in Fig. 5. Thus, the kinetics equation from austenite to twinned martensite can be acquired as

$$\xi_{TM} = \frac{(1-\xi_{TM0})}{1+e^{\frac{T-T_M}{M}}} + \xi_{TM0} \tag{13}$$

where TM is the characteristic temperature for martensite transformation, and numerically equals to the temperature where the complete martensite transformation goes into the half, and is equivalent to the peak temperature of the DSC exothermic curve approximately; Mis the transformation parameter reflecting the width of the exothermic peak, and numerically equivalent to

$$\frac{A_f - A_s}{2e}$$

Here A_s, A_f denote the start and finish temperatures of martensite transformation defined by the traditional method.

Transformation from twinned martensite to single martensite Transformation condition

Assuming in one-dimensional terms, only the increase of stress can lead to this phase transition from twinned martensite to single martensite. Transformation principle: During the uniaxial tension process, for the transformation from twinned martensite to single martensite, if the none macro press stress occur during unloading process, this unloading will not affect the transformation of the subsequent reloading process. During the phase transition process, taking the twinned martensite as austenitic, this transformation will be similar to the transformation from austenite to single martensite, only with the different phase transition peak temperature, and the transformation parameters and impact coefficient of the stress on the temperature still adopted the values of transformation from austenite to twinned martensite. Therefore, based on the kinetics of the transformation from austenite to twinned martensite, kinetics of the transformation from twinned martensite to single one can be achieved as

$$\xi_{SM} = \frac{1}{1 + e^{\frac{T - \frac{\sigma^i}{C_M'} - T_M'}{M'}}} \tag{14}$$

$$\xi_{TM} = \frac{1}{1 + e^{\frac{\sigma^i - \sigma^i_{crit}}{CM' \bullet M'}}} \tag{15}$$

Where $_i$ denotes the average internal stress between the crystals (unit: MPa); TM'is the transformation characteristic temperature; M' is the transformation parameter; CM'denotes the impact coefficient of stress on transformation temperature; icritdenotes the average internal stress between the crystals where the full de-twinned process occurred on the half (unit: MPa).

Transformation from single martensite to twinned one Transformation condition:

For uniaxial tension process, macro-stress and the temperature will not induce the transformation from single martensite to twinned one. However, under repeated stretch-press loading, the martensites with different directions will change its direction repeatedly, and this process corresponds to the SMA rubber-like plastic phenomenon.

Transformation from single martensite to austenite Transformation condition

Assuming in one-dimensional terms, heating and the decrease of stress can lead to this phase transition from single martensite to austenite. Transformation principle: As similar to the transformation from twinned martensite to austenite, for the transformation from single martensite to austenite, if the reverse transformation from austenite to single martensite did not occur during the process, a small cooling or increase of stress will not affect its subsequent transformation process. However, if the reverse transformation process happened and induced a lot of austenitic to single martensite, then the extent of the following transformation from single martensite to austenite required reduction in accordance with the quality percentage of the current single martensite. Thus, the kinetics equation from single martensite to austenite can be acquired as

$$\xi_{SM} = \frac{\xi SM0}{1+e^{\frac{T-\frac{\sigma^i}{C_A}-TA}{A}}} \tag{16}$$

where SM0 is the initial quality fraction of single martensite.

Transformation from austenite to single martensite Transformation condition:

Assuming in one-dimensional terms, cooling and increase of stress can lead to this phase transition from austenite to single martensite. Transformation principle: As similar to the transformation from austenite to twinned martensite, for the transformation from austenite to single martensite, if the reverse transformation from single martensite to austenite did not occur during the process, a small heating or decrease of stress will not affect its subsequent transformation process. However, if the reverse transformation process happened and induced a lot of single martensite to austenitic, then the extent of the following transformation from austenite to single martensite required reduction in accordance with the quality percentage of the current austenite. Thus, the kinetics equation from austenite to single martensite can be acquired as

$$\xi_{SM} = \frac{(1-\xi SM0)}{1+e^{\frac{T-\frac{\sigma^i}{C_M}-T_M}{M}}} + \xi SM0 \tag{17}$$

Parameter variation properties of the kinetics

At the same time, DSC experiments showed that, even for the same material components and the same diameter of the SMA, the transformation of the material would be very different after experiencing loading and plastic deformation, as shown in Fig. 8, and these changes must be considered in the constitutive model. For the SMA of 1mm in diameter, in austenite transformation process, after undergoing the plastic deformation, the transformation peak temperature changes from 50.25 °°C to 121.0 °C, and the transformation parameter changes from 3.25 to 10.25. Similarly, in the martensite transformation process, the transformation peak temperature changes from 27.55 °C to 45.9 °C, and the transformation parameter changes from 1.93 to 11.28. It must be pointed out that these variations are apt for the SMA with 1 mm in diameter. For the SMA with 5 mm in diameter, the change scope of these parameters will be different and can be determined by experiment.

Consideration of the Plastic Deformation

In the first loading process, after stretching to different deformations and heating under free condition, plastic deformations for different initial deformations are shown in Fig. 19. Experimental results indicated that, when the maximum tensile strain exceeded 3%, the plastic deformation increment would be linearly related to the maximum tensile strain increment, thus, the experience equation of the plastic evolution could be achieved as follows

$$\varepsilon^{P} = 0.198 \times (\varepsilon - 0.03) and \varepsilon^{P} \geq 0 \tag{18}$$

It must be pointed out that, this equation is the fitted results of the uniaxial tensile test at room temperature. For another SMA material under different temperature, variation of the plastic deformation with the strain will be very different. Additionally, this equation is only applicable to simulate plastic evolution characteristics during the first loading process, but not the second and the third loading processes.

Numerical Simulation and Comparison with the Experimental Results

Two parts are included in this section, one is the determination of the parameters in the kinetics equation based on the DSC test results and the calculation of the phase transformation, uniaxial tension, SME and the constrained recovery curves, the other is the comparison with the experimental results.

Determination of the Parameters in Constitutive Equation

The transformation characteristics was calculated through the kinetics equation directly and compared with the experimental results. The SMA uniaxial tension, SME and the constrained recovery stress-strain-temperature curves were calculated through the incremental form of the constitutive model. The Parameters used in the calculating process were determined through the following principles: Transformation characteristic temperature and parameters of the kinetics were determined through the DSC test results; The macro physical parameters, such as the elastic modulus, thermal expansion coefficient adopted the experimental result; Because the characteristics temperature and transformation parameters change with the loading process, therefore, in the calculating program, these parameters were adjusted according to the test results of SMA with 1 mm in diameter; The critical value of the internal stress was adjusted based on the experimental critical stress in uniaxial tension.

Verification of SMA Transformation Property

According to the variation curve of martensite quality percentage with the temperature of martensite transformation and reverse one (Fig. 3), the parameters of the kinetics equation can be achieved through the numerical fitting of the test results, as shown in Table 1.

Table 1: Transformation parameters determined by DSC test

	Transformation characteristics temperature (°C)	Transformation parameter (°C)
Martensite transformation process	28.2	1.947
Reverse martensite transformation process	50.14	2.999

Using the parameters calibrated by the numerical fitting, variation curve of martensite quality percentage with temperature can be calculated. The results with comparison of the test ones are shown in Fig. 21. Additionally, variation curve of martensite quality percentage changing rate with temperature can be calculated as shown in Fig. 22, with comparison of the test ones. These comparisons between the numerical and the experimental results indicate that the kinetics equation proposed can simulate the transformation properties of SMA more practically, there are not obvious start and finish temperatures in the calculated curve during the transformation process, the calculated exothermic and endothermic peaks are consistent well with the experimental ones.

Comparison of the Uniaxial Tension, SME and Controlled Recovery Curve

Through the constitutive model with differential form, the uniaxial tension, SME and controlled recovery curve of SMA with 5 mm in diameter can be calculated by program, and compared with the test results, as shown in Fig.s 23-25.

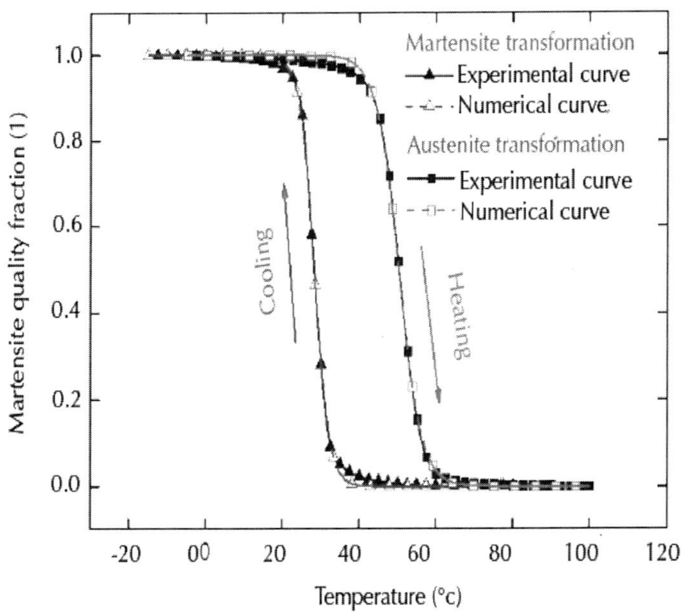

Figure 21: Comparison between experimental and calculated curve of martensite quality fraction versus temperature.

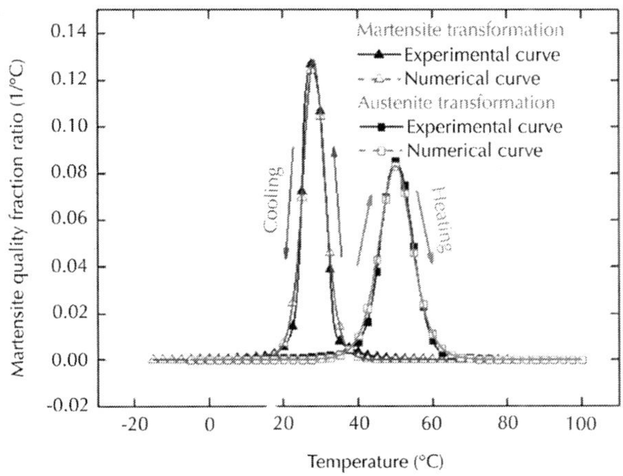

Figure 22: Comparison between experimental and calculated curve of martensite quality fraction ratio versus temperature.

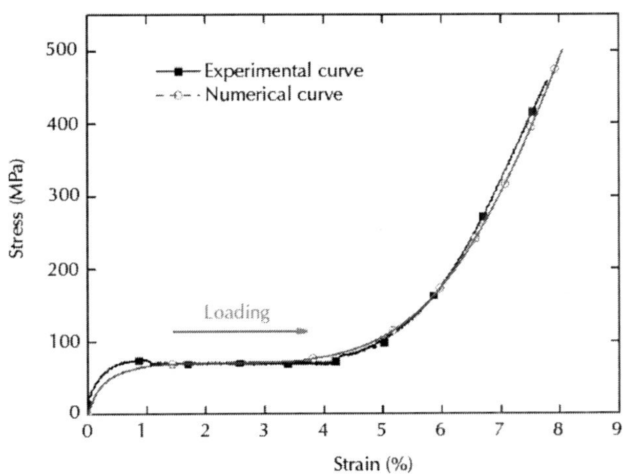

Figure 23: Comparison between experimental and calculated curve of uniaxial tension process.

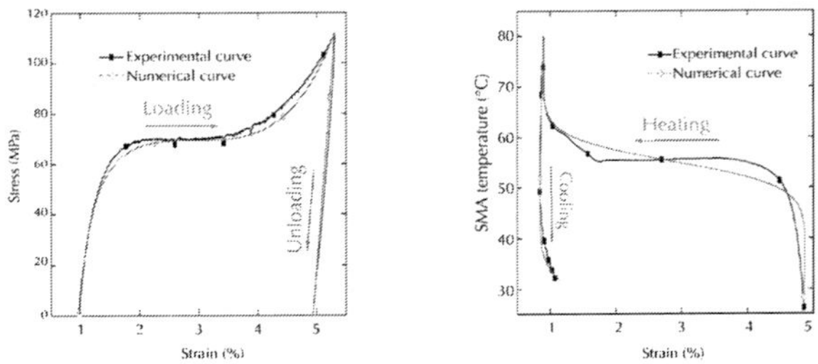

(a).Stress-strain curve

(b). Temperature-strain curve

Figure 24: Comparison between experimental and calculated curve of the third times SME process.

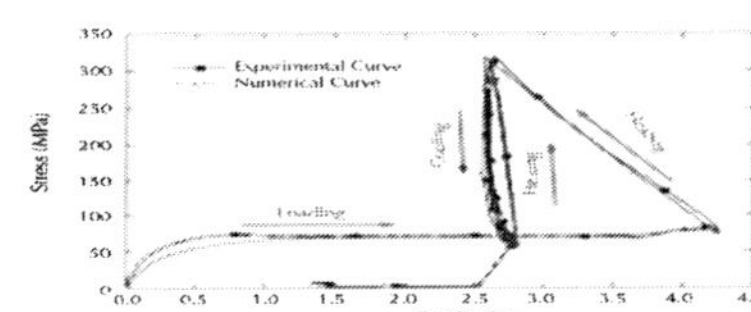

(a). Stress-strain curve

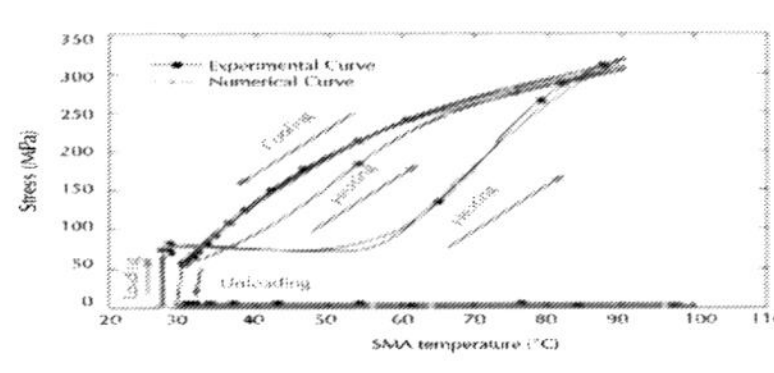

(b).Stress-temperature curve

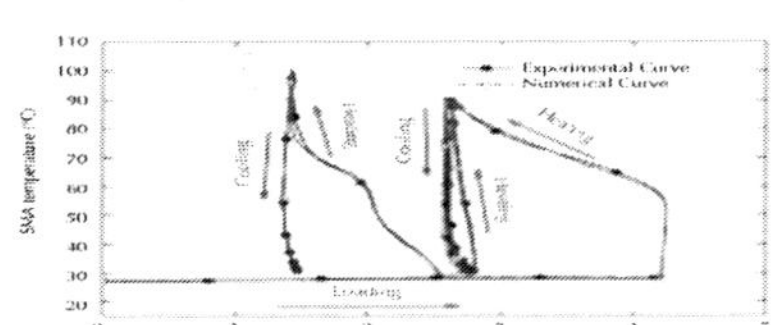

(c). Temperature-strain curve

Figure 25: Comparison between experimental and calculated curve of constrained recovery process.

Numerical results indicate that, based on the well forecast for the phase transformation properties, the constitutive model can simulate the uniaxial tension, SME and controlled recovery process of SMA more accurately.

CONCLUSIONS

A constitutive model in strain form with a new simple empirical kinetics equation is presented based on the DSC and uniaxial tension test, and the plastic deformation is considered. Transformation temperature parameters that denote the peak temperature and peak width of the endothermic and exothermic process were adopted to simulate the transformation characteristics of SMA more practically. Transformation conditions and properties of the six transformation processes were analyzed. At the same time, physical and kinetics parameters were identified through the DSC and the uniaxial test analysis for SMA, programs were adopted for calculating the phase transformation and uniaxial tension process. Comparison between the simulating and experimental results shows that the established constitutive model can simulate the martensite and the reverse transformation, as well as the uniaxial tension, SME and constraint recovery process, more accurately.

Behavior of Smart Concrete Beams with Embedded SMA Bundles

The behavior of smart concrete beams with embedded SMA bundles was investigated. Two beams measuring 1996cm×99cm×85cm, which will be integrated into a smart bridge in a freeway, were manufactured and examined. Each beam contained six trusses of SMA bundles used as actuators to achieve recovery force. The SMA bundles were connected with pre-stressing steel strands and separated from the concrete matrix, so that the temperature interchange between SMA bundles and the matrix could be decreased as small as possible. Some temperature sensors, reinforcement meters and displacement sensors were used to monitor the active control effect of SMA bundles, all the data were acquired through a 16-channel dynamic data-acquisition system, and each beam was examined several times with different activating current intensity. Experimental results indicated that the recovery force induced by SMA bundles was significant and controllable, the deflection generated by the SMA bundles at the middle span of the beam was about 0.44mm, and the capability of resisting overload of each beam was about 2.98 KN (average). A relationship between SMA temperature and activating / inactivating time was also formulated. The conclusion is that SMA could be used in civil engineering structures either from technological or economic aspect.

Background

With the developments of materials science, engineering technology, data acquisition and computer technology, 'Intelligent' materials, such as Electrorheological fluid, piezoelectric ceramic, shape memory alloys, etc, have opened the door for many useful applications, e.g. civil engineering, automotive industry, aerospace technology, medical science (Crawley, 1994; Matsuzaki, 1997;Srinivasan & Mcfarland, 2001). Implementation of smart structures has become feasible, and much effort has been made in these fields.

SMA Smart Structure

Because of the merits of SMA, many researchers have been trying their efforts to apply it in smart structures. These applications were mainly based on the two properties of SMA: one is the shape change with temperature, where SMA fibers are pre-strained during the curing process and prevented from a transformation to austenite, if the temperature rises during the operational cycle, the fibers transform into austenite and tend to contract, and the recovery tensile stresses develop due to the constraint of the matrix; The other is the mechanical characters variation, where unstrained SMA fibers are embedded within the composite, increase or decrease of the temperature alters the elastic modulus or the damping property of SMA, consequently changing the property of the composite. Due to these properties, SMA is usually used to generate bending (Chaudhry & Rogers, 1991; Lagoudas & Tadjbakhsh, 1992), control bucking and postbucking (Baz & Tampa, 1989; Choi et al., 1999, 2000), induce or depress vibration (Baz, Imam, & Mccoy, 1990; Anders, Rogers & Fuller, 1990; Baz, Poh, & Gilheany, 1995; Lau, Zhou, & Tao, 2002), isolate seism (Graesser & Cozzarelli,

1991). Jonnalagadda, Kline, & Sottos (1997) investigated the interaction between SMA wires and a host polymer matrix by correlating local displacements and stress fields induced by the embedded wires with SMA/polymer adhesion, interfacial bond strength was measured for four different SMA surface treatments: untreated, acid etched, hand sanded and sandblasted. Song, Kelly, & Agrawal (2000) presented the design and experimental results of the active position control of a SMA wire actuated composite beam, which has a honeycomb structure with SMA wire embedded in one of its face sheets, a robust controller was designed and implemented to actively control the tip position of the composite beam.

SMA-Based Concrete Structures

In the recent ten years, many researches have been done for the application of these intelligent materials in civil engineering. Maji & Negret (1998) carried out their laboratory studies on concrete (mortar) beams (30.5cm×2.5cm×1.3cm), where SME in NiTi was utilized as a way of inducing additional pre-stressing in concrete. Strands made with NiTi SMA (4 0.64mm) were elongated beyond their plastic limit and subsequently embedded in concrete beams, upon electrical heating, the SMA strands shrunk and induced deflection and failure in concrete. In this study, the temperature rise of the beam was formulated based on heat transfer theory, assuming that the temperature of SMA was the average temperature of the beam. Deng et al. (2003) studied the behavior of concrete specimens uniaxially embedded with shape memory alloy wires actuated by electrical current. Two kinds of specimens with different number of SMA wires were studied, the diameter of the SMA employed in the matrix was 3.5mm and the pre-strains were 8% and 6%. Many factors affecting the behavior of concrete specimens were examined through experiments, such as actuating electrical current mode, initial pre-strain of SMA, actuation times, and initial ambient temperature. The experimental results indicated that the axial strain could be adjusted easily by changing the value of electrical current intensity or the actuating time of the SMA actuator. However, the application of smart concrete beams with embedded shape memory alloy bundles has not been reported to date. The above researches were mostly carried out in laboratory, but a potential practical use of smart concrete beams needs an adequate study on them, because the characteristics of small specimens, as well as the manufacture process, may be very different with those of the engineering structures, thus it is necessary to make a practical engineering investigation in civil structures. Accordingly, two concrete beams measuring 1996cm×99cm×85cm were manufactured and investigated in this section.

EXPERIMENTAL PROGRAM

Materials

The materials used for the host concrete were type I Portland cement, crushed limestone, and river sand. A high-rang water reducer was used to achieve good workability. The mix proportion by weight was cement: sand: aggregate: water: admixture = 1: 1.187: 2.412: 0.329: 0.008. The compressive strength of the concrete (cubic specimen) for 28 days was 53.47 MPa. The seven-wire high-strength steel strands with the diameter of 15.24mm were used to apply conventional pre-stressing force in this experiment, its tensile strength is 1860 MPa, the working stress was controlled to 1395 MPa, and the modulus is 180 GPa. As ordinary reinforced concrete beam, steel reinforcement was used to construct the reinforcing cage. The Ti-49.5 wt % Ni for our research (binary, straight annealed) was one-way SMA. Its transformation temperatures were determined as 25.3 °C (Mf), 36.3 °C (Ms), 44.8 °C (As) and 66.4 °C (Af) through a DSC (differential scanning calorimetry) analysis (Fig. 26). The specific stress-strain behavior of this SMA bar under different temperatures and constrained condition were, as shown in Fig. 27, achieved through a sophisticated test method with a 10KN universal material testing machine. The maximum recovery strain was larger than 5%, and the tensile strength was about 940 MPa (corresponding to the ambient temperature: 22.3 °C) with the ultimate strain of 19.1% (Fig. 27a). When the SMA is free to deform at a relative lower temperature, large plastic strain (about 4%) is induced, but upon heating, this plastic strain disappears, with only a small unrecovery strain (smaller than 0.2%) in the end (Fig. 27b). Whereas, if the SMA recovery is constrained, large stresses should be induced (Fig. 27c), this behavior is the principle of SMA used as actuators.

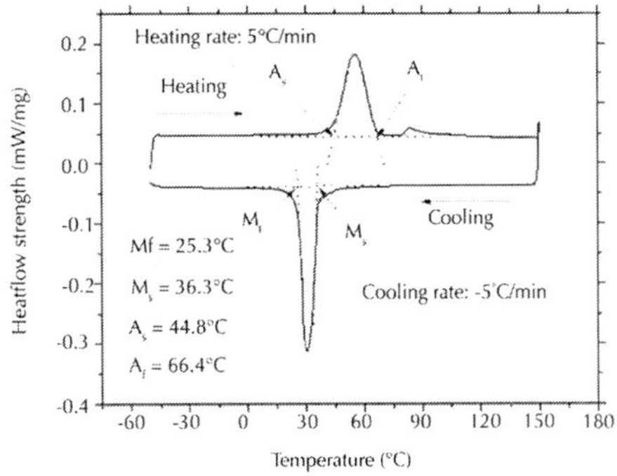

Figure 26: DSC thermogram for NiTi SMA.

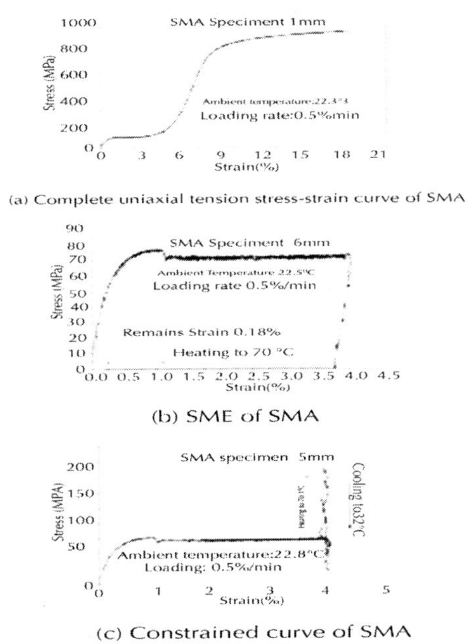

Figure 27: Stress-strain behavior of NiTi SMA.

As depicted in Fig. 28, a seven-wire SMA bundle with dimension of 15.3mm was presented to obtain large recovery force, where one SMA wire with diameter of 5.3mm was surrounded by six SMA wires with diameter of 5.0 mm to obtain good clamping capability.

Smart Concrete Beam Manufacture

The smart concrete beam with dimensions 1996cm×99cm×85cm was manufactured, and the cross section is shown in Fig. 29. Sixteen bounds of pre-stressed high-strength steel strands were mounted on the bottom of the beam to resist the normal loads; six bounds of SMA bundle-steel strand union body, numbered 1 to 6 from left to right, respectively, were constructed upside the steel strands to create recovery forces.

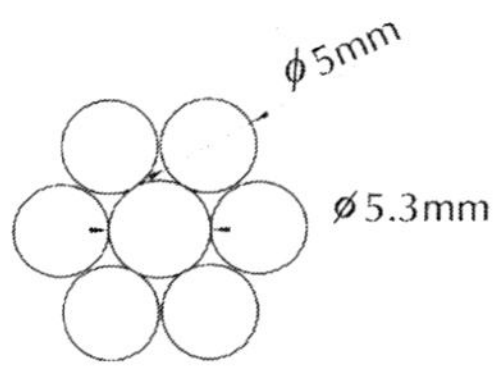

Figure 28: Diagram of SMA bundle.

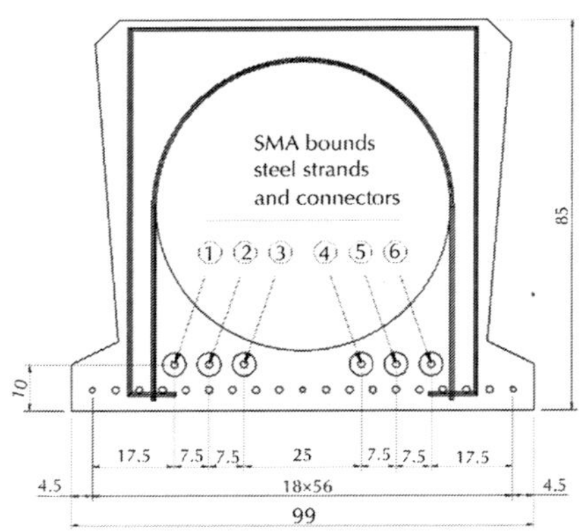

Figure 29: Distribution of cross-section of the smart concrete beam.

A new SMA embedding method in concrete was presented, which can be seen in Fig. 30. The SMA bundle measuring 3.5 m long was mounted in the middle of the beam to create recovery forces, and connector was used to connect the SMA bundle with the steel strands. The recovery forces generated by the SMA bundles were transferred to the concrete matrix by the steel strands on the bonding zone measuring 3 m at both ends of the beam. When the SMA bundles were activated by electricity, additional bending force was generated and applied to the beam, In the unbonded zone, SMA bundles and steel strands were separated from concrete by PVC pipes, therefore the SMA bundles and the connectors could shift freely, and the temperature influence of SMA bundles to the matrix could be decreased as small as possible.

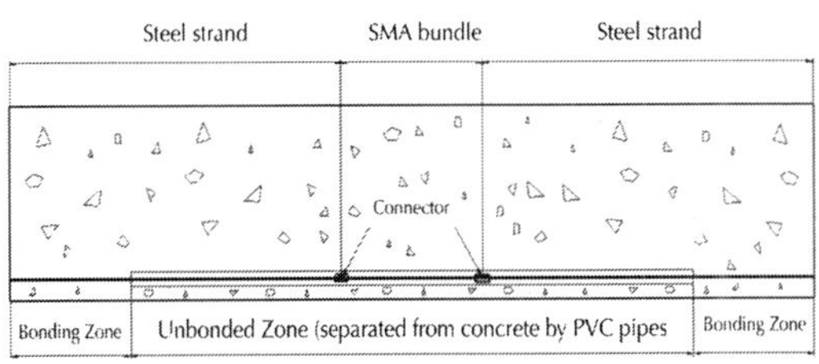

Figure 30: Distribution of longitudinal-section of the smart concrete beam.

Sensing and power system of these smart beams can be seen in Fig. 31. Temperature sensors and reinforcement meters were mounted on the SMA bundle, six trusses of SMA bundles were connected in series by voltaic wires to receive the electric power, and copper wiring terminals were used to connect the conductor wires with the SMA bundles. All the sensor wires were laid along the beam at a lower level position and led out at the end of the beam. Two voltaic wires were laid along the beam at a higher level position and led out at the same end of the beam. Considering the effect of steel reinforcement cage, the electric circuit must be constructed as a self-consistent system to avoid short circuit, and all sensor wires and electric cable should be protected from the damage of the construction machinery.

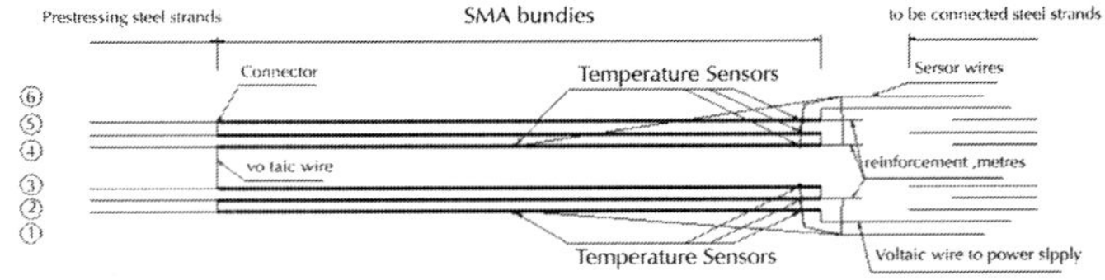

Figure 31: Layout of the sensors and wires.

Initial tension of the SMA bundles was carried out after the initial elongation of the steel strands, and the initial strain of the SMA bundles reached 4%. After finishing the stretching of the SMA bundles and steel strands (Fig. 32), concrete was deposited into the steel model and steam was used during the curing period of the concrete.

Figure 32: Framework of steel reinforcement with stretched SMA bundles.

After about three days, when the concrete strength reached 80 percent of its 28–day strength, SMA bundles and steel strands were relaxed and the concrete beams were lifted out from the model, and tests of these beams were carried out subsequently.

Testing Apparatus and Procedure

The beam to be monitored was laid on an experimental bench for freely-supported condition, and the testing site is shown in Fig. 33. When the SMA bundles were activated, additional bending forces were generated, and the beam was inflected at the same time. Contrarily, when the SMA bundles were inactivated, the recovery forces disappeared and the beam recurred to the original state, as shown in Fig. 34.

Figure 33: Experiment site.

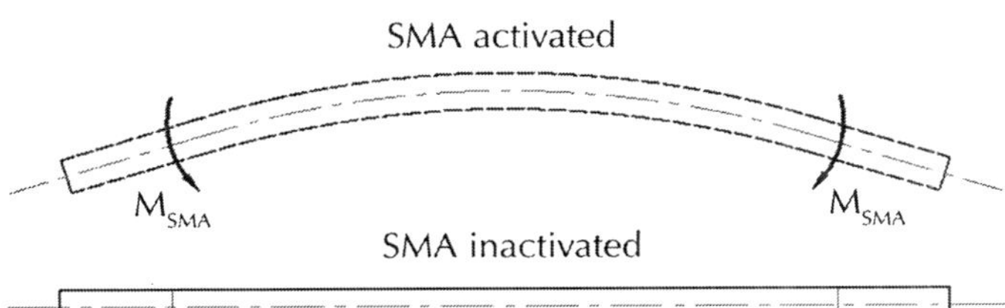

Figure 34: Behavior of the beam with SMA activated and inactivated.

The SMA bundles were heated with electricity. Because of the low electrical resistance of the SMA bundles, the necessary voltage was low, but the required current capacity was large, therefore, a 12.5V, 125A variable capacity voltage source was employed to each beam. Experimental results indicated that the resistance of the SMA bundles in each beam was about 0.166Ω. Three displacement sensors were mounted at three quartiles of the beam. Deflection of the beam, the temperatures of the SMA bundle, and the recovery forces generated by the SMA bundles, were monitored synchronously during the progress of the experiment. All the data were acquired and analyzed preliminarily through a 16-channel dynamic data-acquisition system. The sampling frequency was 10 Hz, and the maximum sampling frequency could reach 10000Hz.

RESULTS AND DISCUSSIONS

Two beams were manufactured and each was examined for several times. For convenience, beams, SMA bundles and temperature sensors were all numbered by the following principle: Two beams were numbered 1 to 2. Serial number of the SMA bundles of each beam was composed of the number of the beam and relative position label in the same beam, i.e. 2-4. The first number indicated that the SMA bundles belonged to the No. 2 beam and the second number indicated that this is the No. 4 SMA bundle, as depicted in Fig. 29. Temperature sensors were marked by the number of the SMA bundle and the position label, i.e. 1-3-T1. The first number of the temperature sensor label indicates the number of the beam, and the second denotes the number of the SMA bundle where the temperature sensors are mounted on. The symbol T1 denotes that the temperature sensors are mounted in the middle of the SMA bundles, and T2 denotes that the temperature sensors are mounted at the end of the SMA bundles. Thus the flag 1-3-T1 indicates the temperature sensor mounted in the middle of 1-3 SMA bundle.

Deflection of the Smart Concrete Beam

The recovery forces generated by the SMA bundles would induce a bending force on the beam, and cause a deformation of the beam, as depicted in Fig. 34. The deformation of the beam was indicated by the displacement of the three quartiles on the beam. The effect of the environmental temperature, which was discussed in the following content, was eliminated during this analytic process. Each beam was activated and monitored for several times, and the results were almost similar. The representative results of beam 1 and beam 2 were shown in Fig. 35.

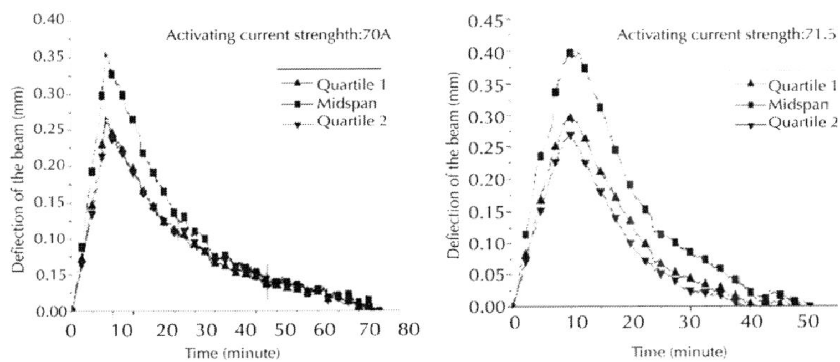

(a). Beam 1 in no.3 activating times (b). Beam 2in No.9 activating times

Figure 35: Deflection of the beam versus time.

Temperature of SMA Bundles

When the beam was activated with voltage, temperature of the SMA bundles would rise, and when the current source was shut down, the temperature would drop. The maximum temperature rising speed was at the initial point, as time went on, it decreased and approached to zero, then the thermo-genesis was equal to the heat dissipation and temperature kept up a constant maximum value. Different temperature sensors mounted on the different SMA bundles in a beam detected different temperatures, as shown in Fig. 36. This phenomenon came from the different heat conducting boundary condition of the different SMA bundles. But the results of the same temperature sensor under different activating times were little different, as depicted in Fig. 37, only the temperature dropping section had a little difference yielding from different temperature environment in different activating times.

Recovery Force Generated by SMA Bundles with Temperature

When the SMA bundles were activated by voltage, the temperature of these bundles raised and the recovery forces generated and increased, but when the current was cut off, the temperature dropped and the forces decreased. The different force generated by different SMA bundles, this difference yield from the different initial force of the SMA bundles. Force variation with time of some bundles was shown in Fig. 38, and each symbol indicated 1.5 minutes. The recovery forces almost increased proportionally with the rising of the temperature, just as depicted in Fig. 39. Hysteresis cycles were observed in the curve of force versus temperature of SMA bundle 2-4, this phenomenon will be investigated later.

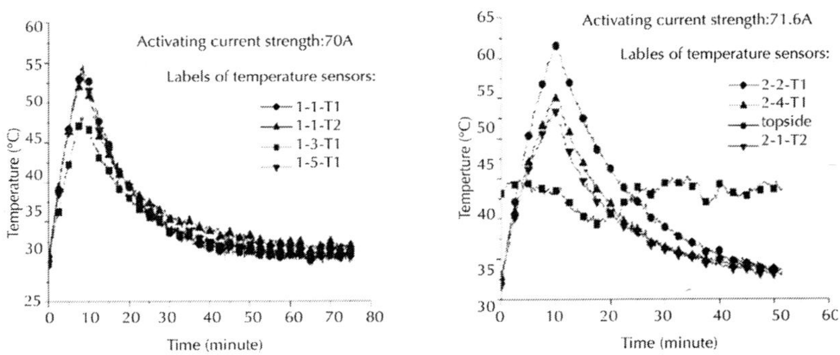

(a). Temperature sensors of beam 1 in NO.3 activating times

(b). Temperature Sensors of beam 2 in No.9 activating times

Figure 36: Temperature versus time for different temperature sensors in the same beam.

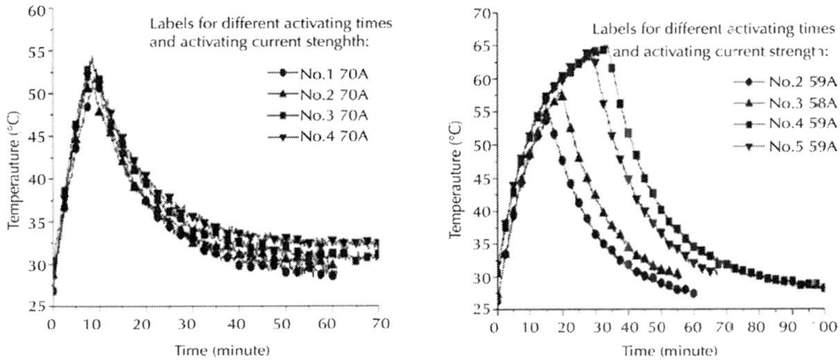

(a). Temperature sensor 1-5-T1 (b). Temperature sensor 2-2-T1

Figure 37: Temperature versus time for different activating times of the same temperature sensor.

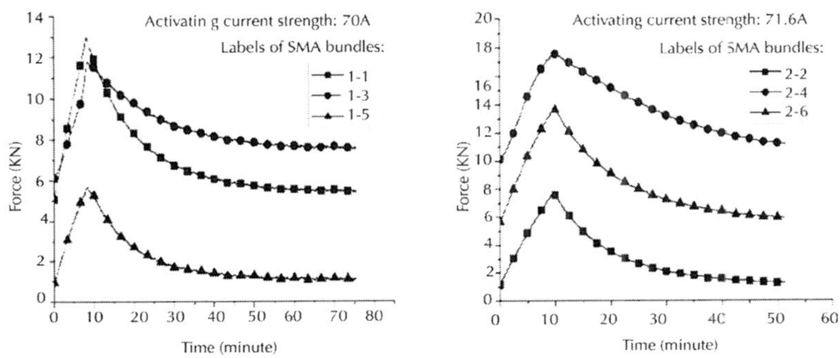

(a). SMA bundies of bham 1 in No.3 activating times

(b). SMA bundies of beam 2 in No.9 activating

Figure 38: Force versus time for different SMA bundles of the same beam.

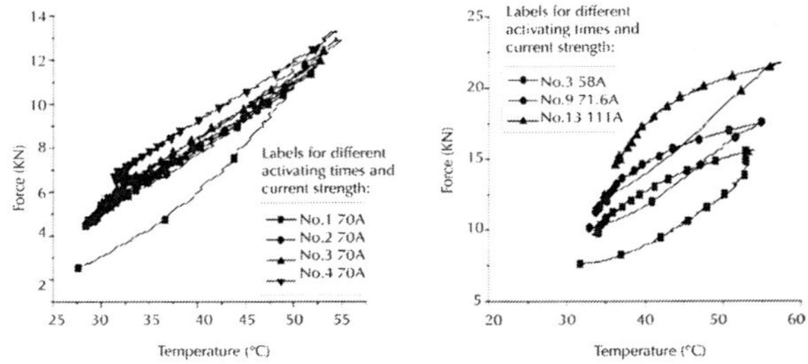

(a). SMA bundle 1-1 (b). SMA bundle 2-4

Figure 39: Force versus temperature for different activating times of the same SMA bundles.

Effect of Activating Current

Effect of the activating current was investigated, current with the strength of 58A, 70A and 110A were applied to the SMA bundles, higher current would induce a faster rising of the temperature of the SMA bundles, thus, the recovery forces increased faster, and the beam reacted with a higher speed, but the recover processes were almost similar, just as depicted in Fig. 40, 41 and 42.

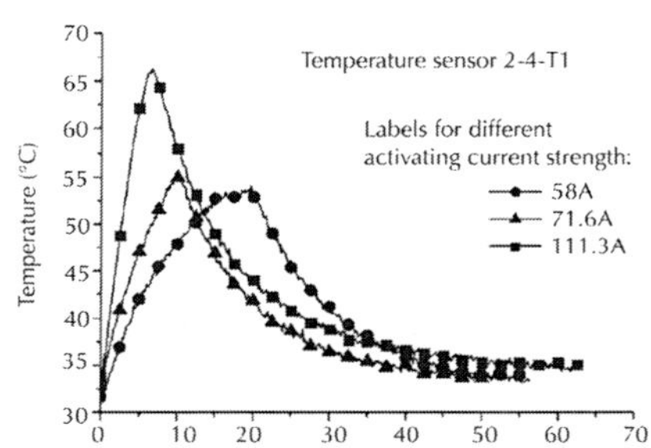

Figure 40: Temperature versus time of temperature sensor 2-4-T1 for different activating current.

Effect of Environment Temperature

During the experiment, the upside temperature of the beam would increase because of the sunshine, while that of the lower side varied rarely. Therefore, this uneven temperature distribution would induce the deflection of the beam. Deformation and the variation of the temperature of the beam also monitored to explain this effect on the deflection of the beam, and the result of this effect was depicted in Fig. 43. Monitoring of No.1 beam was carried out at 9:00 AM and finished at 8:00 PM. The maximum temperature on the upside of the beam (T0) occurred at 3:00 PM, but the maximum deflection the beam at 5:40 PM, this time lag showed that it needed heat conductivity from the surface to the inner. Temperature on the lower side of the beam varied rarely. In the morning, right-hand of the beam faced the sunshine, and in the afternoon, the left-hand of the beam, therefore, temperature at the right-hand of the beam (T2 and T3) increased faster than that of the left-hand of the beam (T1) in the forenoon, and slower in the afternoon, just as depicted in Fig. 43. During this monitoring course, all the SMA bundles were inactivated, thus the temperatures of the SMA bundles were equal to these of the concrete matrix. From this result, it can be seen that the effect of the environment temperature is considerable.

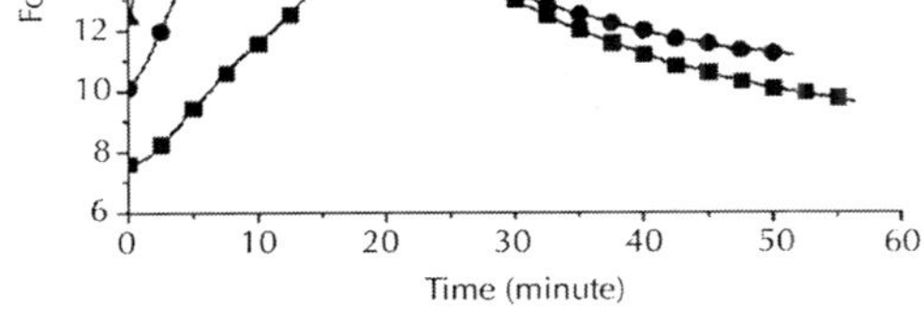

Figure 41: Force versus time of SMA bundle 2-4 of different activating current.

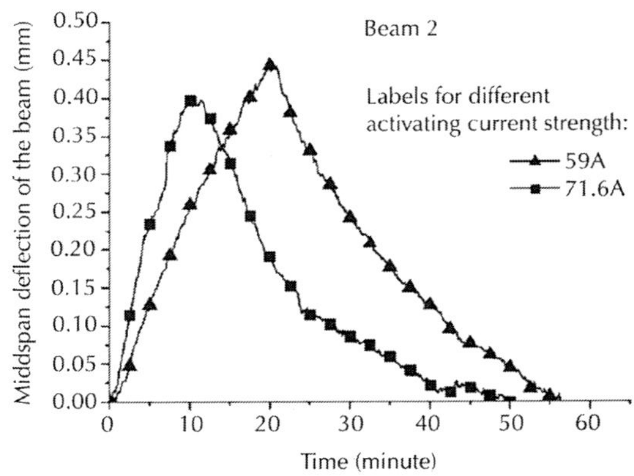

Figure 42: Middle span deflection versus time of beam 2 for different activating current.

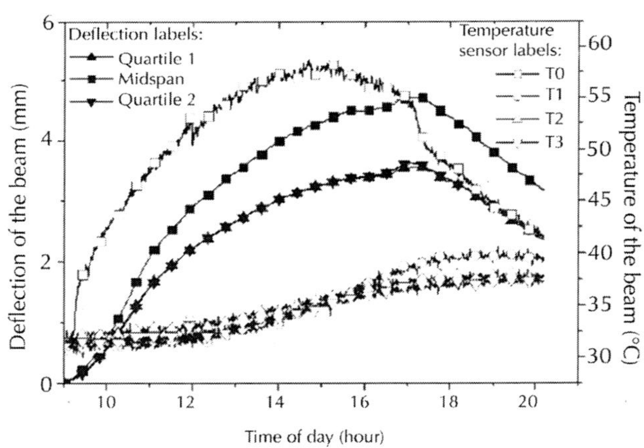

Figure 43: Effect of environment temperature on deflection and temperature of the beam.

Theoretical Analysis

Resisting Overload Capability

Assuming that a short-term concentrated overload P was applied to the beam, as depicted in Fig. 44, then the overload inducing 1 mm displacement at the middle span of the beam could be calculated.

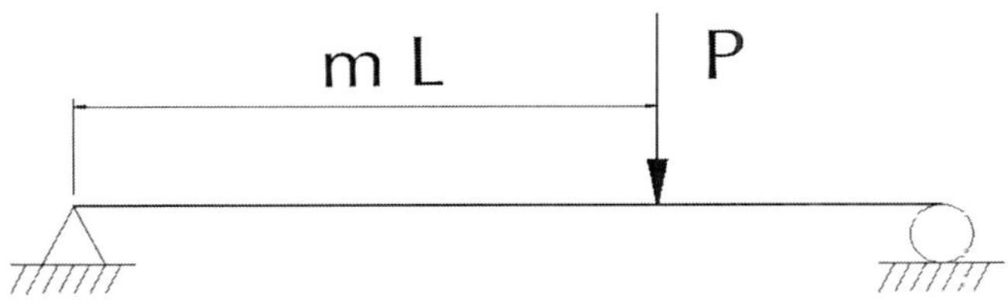

Figure 44: Schematic of load application model

When the overload was applied to the middle span of the beam, the maximum flexural torque at the middle span of the beam generated by the overload was:

$$M_{max} = \frac{PL}{4} \tag{19}$$

Where P is the overload, and L is the span of the beam. Thus, the deflection at the middle span of the beam could be given as:

$$f\max = \frac{1}{12}\frac{M_{max}L^2}{B_{s1}} \tag{20}$$

Where fmax is the maximum deflection of the beam (at the middle span); L is the span of the beam measuring 1996cm; Bs1 is the flexural rigidity of the beam for short-term load; Mmax is the maximum flexural torque at the middle span of the beam. Thus, the maximum concentrated overload inducing 1 mm displacement at the middle span of the beam could be calculated as:

$$P = \frac{48 f_{\max} B_{s1}}{L^3} = 6.73KN \tag{21}$$

Assuming that value of the overload inducing a deflection of 1 mm at the middle span of the beam is equal to that of the resisting force in the negative direction at the middle span of the beam, thus the statistic result of the deflections at the middle span of the beam and the resisting force generated by the SMA bundles for each beam could be seen in table 2.

Table 2: Statistic result of deflections at the middle span of the beam and equivalent resisting force of the beams

Beam serial number	The maximum deflection at the middle span of the beam (unit: mm)	Equivalent resisting force (unit: KN)	Activating current strength (unit: A)	Activating time (unit: minute)
No. 1	0.402	2.7055	70	9.3
No. 2	0.482	3.2439	71	16.3

Curve Fitting of Experimental Results

Considering a section of SMA bundle, if T is defined as the temperature of the SMA bundle, then the increase of T is given as

$$\rho c V_{SMA} \frac{dT}{d\tau} = q_v V_{SMA} - a\mu F_{SMA}(T - t_f) \tag{22}$$

Where , c , and VSMA denote density, specific heat, time and capacity of the SMA bundle; qv is the rate of heat production; a , μ , F_{SMA} and t_f denote the coefficient of heat convection, quotient of contact face form, area of the heat-exchange surface and ambient temperature. The rate of heat production $q_v V_{SMA}$ can also be expressed as follows:

$$q_v V_{SMA} = I^2 R_{SMA} \tag{23}$$

Where I and R_{SMA} denote the current strength and electric resistance of SMA. Thus, equation (4) could be written as:

$$\frac{dT}{d\tau} + \frac{a\mu F_{SMA}}{\rho c V_{SMA}} t = \frac{I^2 R_{SMA}}{\rho c V_{SMA}} + \frac{a\mu F_{SMA}}{\rho c V_{SMA}} t_f \tag{24}$$

Defining

$$X = \frac{I^2 R_{SMA}}{\rho c V_{SMA}} \tag{25}$$

$$Y = \frac{a\mu F_{SMA}}{\rho c V_{SMA}} \tag{26}$$

Then the solution to equation (24) could be given as:

$$T(\tau) = Ae^{-Y\tau} + \frac{X}{Y} + t_f \tag{27}$$

Where A is related to the initial condition during temperature rising period. In the same manner, the decrease of T could be given as:

$$\frac{dT}{d\tau} + \frac{\alpha\mu F_{SMA}}{\rho c V_{SMA}} T = \frac{\alpha\mu F_{SMA}}{\rho c V_{SMA}} t_f \tag{28}$$

Thus from equations (26) and (28), we have:

$$T(\tau) = Be^{-Y\tau} + t_f (\text{When } I = 0) \tag{29}$$

Where B is related to the initial condition during temperature dropping period. The experimental curve could be fitted well with this theoretical curve, just as depicted in Fig. 45. Considering equation (22), temperature rising rate of the SMA bundles dT/d could be given as:

$$\frac{dT}{d\tau} = \frac{I^2 R_{SMA}}{\rho c V_{SMA}} + \frac{\alpha\mu F_{SMA}}{\rho c V_{SMA}}\left(t_f - T\right) \tag{30}$$

Assuming that the temperature T was equal to the environment temperature t_f at the beginning, then the initial temperature rate of the SMA bundles dT/d can be given as

$$\frac{dT}{d\tau} = \frac{R_{SMA}}{\rho c V_{SMA}} I^2 \tag{31}$$

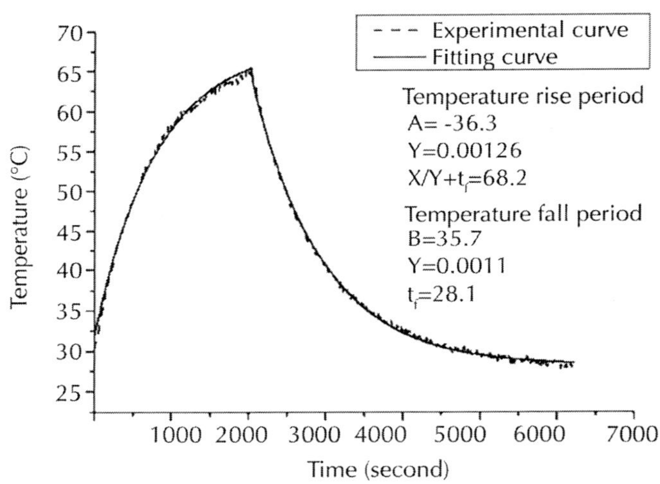

Figure 45: Fitting curve of temperature versus time of temperature sensor 2-2-T1 in NO.4 activating times.

According to the above theory, statistic values of the initial temperature rising rate were fitted with the numerical curve, just as depicted in Fig. 46. From it, one can see that the activating current strength had significant effects on the behavior of the smart beam, and higher activating current could be used to achieve a faster reacting speed of the beam. According to the above analyses, the action speed was the proportion square of the current strength. Theoretically, the temperature of SMA is proportional to the square of the applied current because of this electric heating method. As a matter of fact, this relationship may be affected by many factors, e.g. the heat transfer from the SMA to the concrete matrix. Here, the well accord between the experimental data and the fitting curve indicated that influences of heat transfer between the SMA and the concrete matrix could be decreased effectively by separating SMA bundles from concrete matrix by PVC pipes.

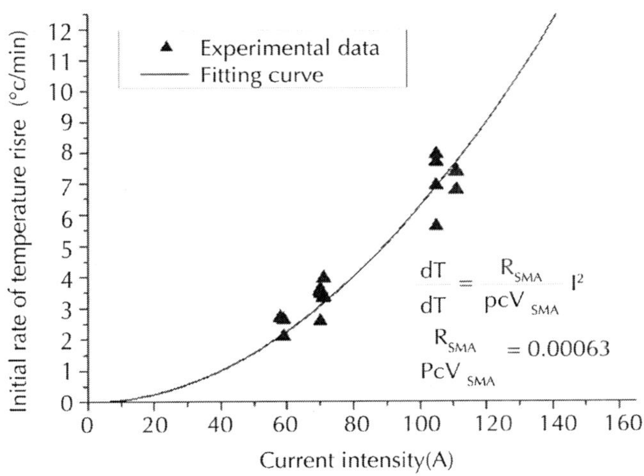

Figure 46: Fitting curve of temperature rising rate versus current intensity.

CONCLUSIONS

In this section, the manufacture process and behavior of the smart concrete beams with embedded SMA bundles were investigated through an extensive experimental program. The main parameters, such as temperature and recovery force of SMA bundles and the deflection of the concrete beam, were monitored through this program. Major factors influencing the behavior of the beams were examined. From these results, the following conclusions can be drawn. SMA can be used in practical civil engineering structures to resist overload, the effect is considerable. SMA bundles can change the mechanical performance of the smart concrete beam as needed. In this section, a new SMA embedding method in concrete was presented and applied to this experiment -- only a short section of SMA bundle could generate large force, which was mainly due to the large recoverable deformation of SMA. The recovery force created by the SMA bundles was almost proportional to the temperature of the SMA bundles, and hysteresis cycles were observed in the recovery force-temperature curve of some SMA bundles. The recovery force generated by each SMA bundle was different, which came from the different initial condition for different bundles. But the increments of the recovery force of each SMA bundle in each activating time were almost the same. The temperature of different SMA bundles was different, and temperature of a same SMA bundle in different activating times was also little different. These differences yielded from the different heat conducting condition of different SMA bundles and the different environment boundary condition in different activating times. Due to these differences, more force sensors and temperature sensors would be obligatory for more comprehensive monitoring of the beam. The effect of the environment was remarkable, and displacement at the middle span of the beam induced by the environment temperature was about twelve as times as that caused by the recovery force of SMA bundles. Thus, this effect must be considered during the experimental and analytic process. Some other factors, such as the activating times, the length of the unbonded section, the length of the SMA bundles, the pre-strain of the SMA bundles, and the curing condition of the concrete, etc., can also affect the behavior of the concrete beam. These effects deserve more investigation.

Behavior of a Smart Concrete Bridge with Embedded SMA Bundles

Since there are many differences between the laboratory researches and the practical applications of using this smart material to structural active controlling; thus, it would be worthful and necessary to study how to apply SMA on a practical concrete bridge and how to use the "smart" forces. Accordingly, the objective of this section is to investigate the behavior of a smart concrete bridge with embedded SMA bundles. The bridge was a simply-supported pre-stressed concrete one with 20 meters long and 12.5 meters wide, and composed of ten ordinary concrete beams and two smart concrete ones measuring 1996cm×99cm×85cm. When the SMA bundles were activated, additional bending forces were generated, and the beam was inflected to resist excessive load at the same time. Otherwise when the SMA bundles were inactivated, the smart forces disappeared and the beam recovered to the initial state. In order to monitor the active control effect of the SMA bundles, some temperature sensors, reinforcement meters and displacement sensors were employed. All the data were acquired through a 16-channel dynamic data-acquisition system. The bridge was examined several times with different activating current intensity, and some static load tests were also carried out to evaluate the capability of resisting overload of the bridge. The experimental results indicate that the smart forces induced by SMA bundles were significant and controllable, the deflection generated by the SMA bundles at the middle span of the bridge was about 0.03 mm, and the capability of resisting excessive load of the smart beams was 5.6 ~ 7.5 KN. It is also shown that SMA could be used as active controlling actuator in civil engineering structures.

EXPERIMENTAL PROGRAM

Design Concept of the Smart Concrete Bridge

A complete smart concrete bridge in a freeway should contain the following three parts: load and speed sensors mounted at several miles beyond the bridge; control unit and power supply; actuators, temperature and force sensors, and the bridge matrix, the design concept of the smart bridge was demonstrated in Fig. 47. The two smart beams will be fixed on the smart bridge. In the session of the freeway, load and speed sensors will be amounted at several miles away from the bridge. Therefore, SMA bundles embedded in the smart beams can be activated before the load exerts on it and recovery forces generate to resist the external load. Additionally, the recovery force generated by the SMA bundles is only used to resist the excessive load when needed, normal load is resisted by ordinary structures, such as steel reinforcements and steel strands. After the load passes the bridge, SMA bundles will be inactivated.

Manufacturing the Beams and the Bridge

The bridge was composed of twelve concrete beams, and the two smart concrete beams were fixed on the No. 5 and No. 6 from outer side (Fig. 48). Gaps between the beams were filled up with concrete, and then the surface concrete and the pavement were constructed subsequently.

Testing Apparatus and Procedure

The two smart concrete beams were electrically activated several times to measure the effect of the recovery forces. To evaluate the performance of the bridge, the static load tests were also carried out by parking a 202KN-weight truck on the nine positions of the bridge (Fig. 49).

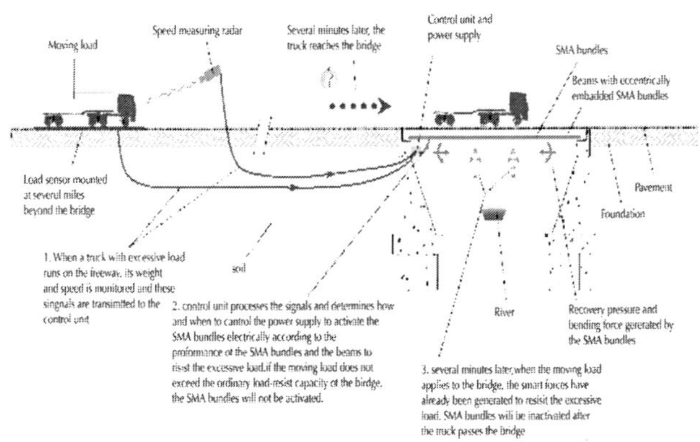

Figure 47: Diagram of the design concept about a smart bridge of a freeway.

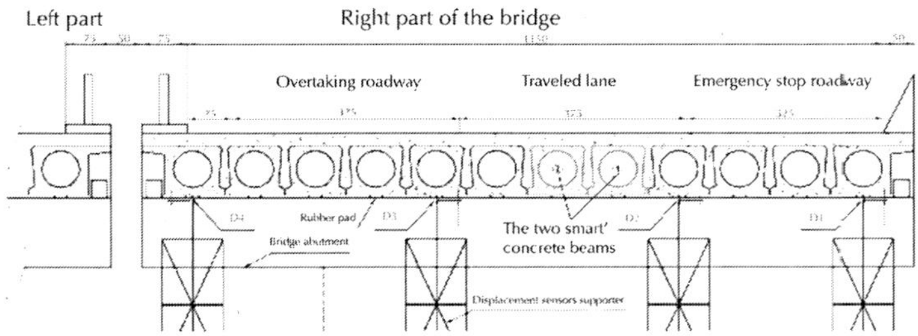

Figure 48: Schematic of mounting deflection sensors.

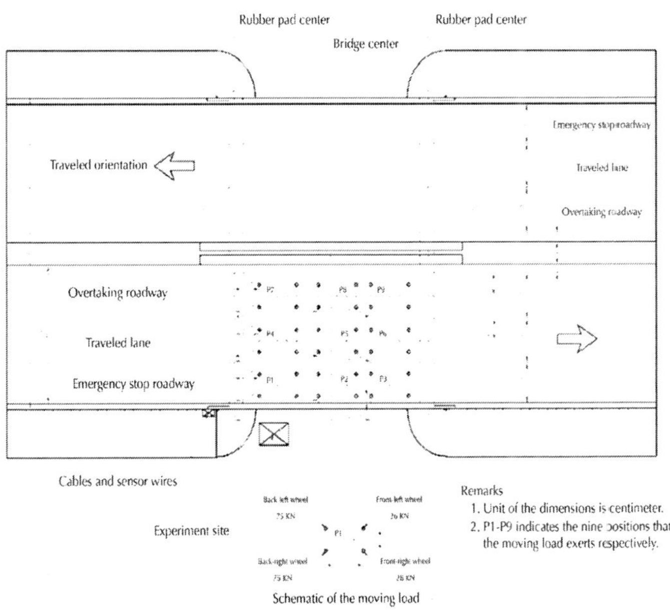

Figure 49: Schematic of static load test.

Results and Discussions

In order to record easily, the SMA bundles, temperature and force sensors were numbered serially. Two beams were numbered 1 and 2, and SMA bundles, 1 to 6. Thus SMA bundles and the force sensors can be indicated with the number of the beam and the relative position label in the same beam. The temperature sensor was denoted via the number of the SMA bundle and its position label (i.e. T1 denotes the temperatures mounted at the middle of the SMA bundle, T2 the end).

Activating Test Results of the Bridge

To evaluate the effect of the recovery force, the SMA bundles of the two smart concrete beams were activated two times without applying any external load on the bridge. Displacement sensors were mounted at the bottom of No 1, 4, 8, 12 beams of the bridge (denoted by D1, D2, D3, D4 respectively) to monitor its deformation, as shown in Fig. 48.The unit of the dimensions is centimeter, and D1, D2, D3, D4 indicates the four positions that the displacement sensors mounted on respectively. Temperatures, as well as recovery forces, were also measured synchronously during the experiment. When the SMA bundles were activated, the bridge was deformed and upward deflections were induced. Deflections at the middle span of the bridge increased and decreased during the activating and inactivating, as shown in Fig. 50. The maximum deflection of the bridge was about 0.03mm. Displacement sensor mounted at the D1 position did not work normally, so the result will be not considered in the analysis. During the heating and cooling process, temperatures of SMA bundles also increased and decreased, as shown in Fig. 51, and recovery forces of SMA bundles increased and decreased proportionally with these temperatures, as shown in Fig. 52. The recovery force created by the SMA bundles was almost proportional to the temperature of the SMA bundles, and hysteretic cycles were observed in the recovery force-temperature curve of the SMA bundles.

Static Test Results of the Bridge

The static load was applied on the nine positions of the bridge, as shown in Fig. 49, deflections at the four representative points of the bridge (indicated as D1, D2, D3, D4) were demonstrated in Fig. 53.

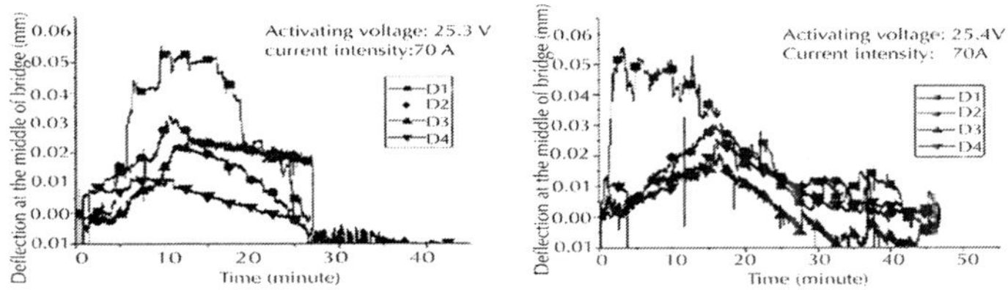

(a) No.1 activating times (b) No.2 activating times

Figure 50: Deflection at the middle of the bridge versus time.

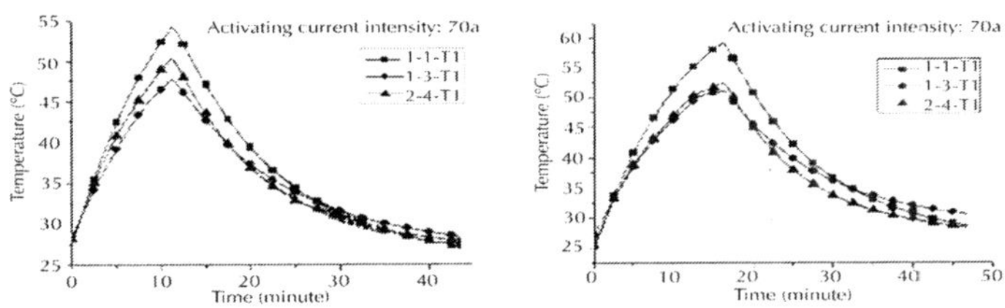

(a) No.1 activating times (b) No.2 activating times

Figure 51: Temperature of SMA bundles versus time.

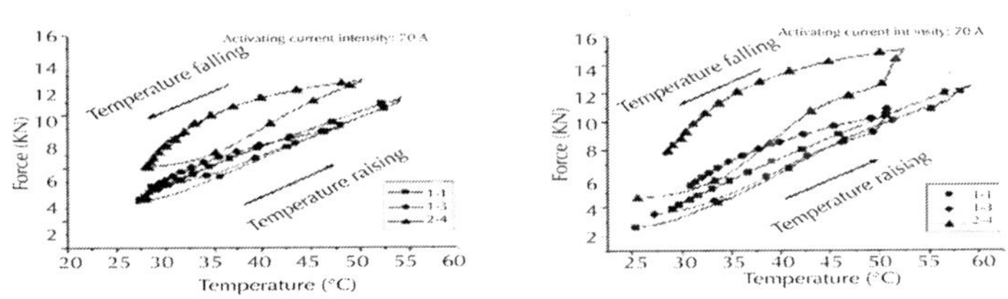

(a) No.1 activating times (b) No.2 activating times

Figure 52: Recovery force versus temperature.

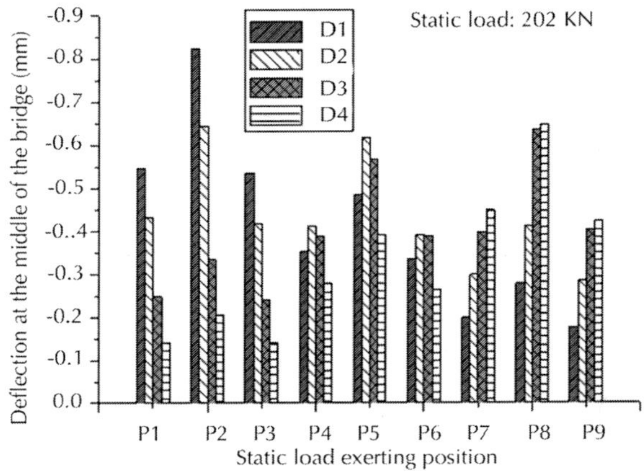

Figure 53: Deflection of the bridge under different load exerting positions.

Effect Analysis of the Recovery Forces

The capability of resisting excessive load of recovery forces was estimated through the following principle: all the deflections at the D1, D2, D3, D4 induced by the recovery forces are larger than or equal to these reverse ones induced by the external load. Therefore, by comparing the average maximum deflections of the activating test and deflections of the static load test, the maximum resisting load capability could be obtained. Different position the excessive load exerted on is, different the resisting load of the SMA bundles. When the load ran on the middle of the traveled lane, deflection of the beam No.8 induced by the recovery forces reached the reverse one by the external load firstly, thus the resisting load of the recovery forces was about 7.53 KN shown in Fig. 54 (a), and on the overtaking roadway, 5.55 KN in Fig. 54 (b).

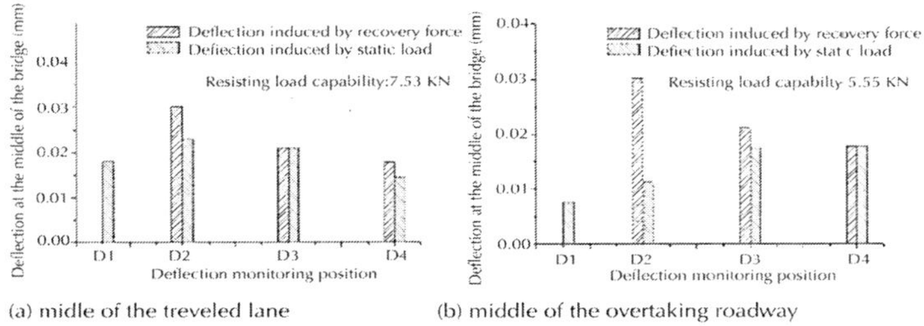

Figure 54: Resisting load analysis.

CONCLUSIONS

In this section, the manufacturing process and the behavior of the smart concrete bridge with embedded SMA bundles were investigated. From these results, the following conclusions can be drawn: (1) SMA can be used in practical concrete bridges to resist external load with considerable effect. SMA bundles can change the mechanical performance of the smart concrete beam as needed; (2)The capability of resisting excessive load of the recovery forces can be varied with the different exerting position of the external load, thus the optimum design of the smart concrete structures will be necessary to obtain the best control effect.

REMARKS

In this chapter, the constitutive characteristics of NiTi SMA and its application in practical concrete structures were investigated. The experimental results indicated that the mechanical properties and the phase transformation characters of NiTi SMA could be affected by the loading process considerably. Comparison between the numerical and experimental results indicated that the proposed model could simulate the phase transformation characters, the uniaxial tension and the constrained recovery behavior of NiTi SMA more effectively. Experimental results indicated that the additional recovery forces generated by the SMA could be changed by altering its temperature, and can be used to adjust the deflection of the concrete bridge and to enhance the load-bearing capability of the concrete bridge.

REFERENCES

1. W. S. Anders, C. A. Rogers, C. R. Fuller, 1990Vibration and low frequency acoustic analysis of piecewise-activated adaptive composite panels," Journal of Composite Materials, 261103120
2. A. Baz, K. Imam, J. Mccoy, 1990Active vibration control of flexible beams using shape memory actuators," Journal of Sound and Vibration, 1403437456
3. A. Baz, S. Poh, J. Gilheany, 1995Control of the natural frequencies of Nitinol-reinforced composite beams," Journal of Sound and Vibration, 1851171183
4. A. Baz, L. Tampa, 1989Active control of buckling of flexible beams," Failure Prevention & Reliability, ASME, 211218
5. L. C. Brinson, 1993One-dimensional constitutive behavior of shape memory alloys: thermomechanical derivation with non-constant material functions and redefined martensite internal variable," Journal of Intelligent Material Systems and Structures, 42229242
6. M. Brocca, L. C. Brinson, Z. P. Bazant, 2002Three-dimensional constitutive model for shape memory alloys based on microplane model," Journal of the Mechanics and Physics of Solids, 50510511077
7. W. J. Buehler, J. V. Gilfrich, R. C. Wiley, 1963Effect of low-temperature phase changes on the mechanical properties of alloys near composition TiNi," J. Appl. Phys., 34514751477
8. Z. Chaudhry, C. A. Rogers, 1991Bending and shape control of beams using SMA actuators," J. Intelligent Mat. Syst. Struct., 24581602
9. Choi Sup, Lee Jung Ju, Seo Dae Cheol, & Choi Sun Woo.1999The active buckling control of laminated composites beams with embedded shape memory alloy wires," Composite Structures, 471-4679686

10. Choi Sup, Lee Jung Ju, Seo Dae Cheol, & Dong Kusong.2000A study on the bucking and postbuckling control of composite beams with embedded NiTi actuators," Journal of Composite Materials, 341714941510

11. Edward. F. Crawley, 1994Intelligent structures for aerospace: a technology overview and assessment," AIAA Journal, 32816891699

12. Deng Zongcai, Li Qingbin, Jiu Anquan, & Li Lei.2003Behavior of Concrete Driven by Uniaxially Embedded Shape Memory Alloy Actuators," Journal of Engineering Mechanics, 1296697703

13. L. Delaey, R. V. Krishnan, H. Tas, H. Warlimont, 1974Thermoelasticity, pesudoelasticity and the memory effects associated with martensitic transformations (part 1 structural and microstructrual changes associated with the transformations)," Journal of Materials Science, 9915211535

14. Funakubo Hiroyasu.Translated from Japanese by Kennedy J.B.) 1987Shape Memory Alloys," Gordon and Breach Science Publishers, New York.

15. A. B. Greninger, V. G. Mooradian, 1938Strain transformation in metastable beta copper-zinc and beta copper-tin alloys," Trans. Met. Soc. AIME, 128337368

16. E. J. Graesser, F. A. Cozzarelli, 1991Shape-memory alloys as new materials for aseismic isolation," ASCE Journal of Engineering Mechanics, 1171125902608

17. A. B. Greninger, V. G. Mooradian, 1938Strain transformation in metastable beta copper-zinc and beta copper-tin alloys," Trans. Met. Soc. AIME. 128337368

18. K. Jonnalagadda, G. E. Kline, N. R. Sottos, 1997Local displacements and load transfer in shape memory alloy composites," Experimental Mechanics, 3717886

19. R. V. Krishnan, L. Delaey, H. Tas, H. Warlimont, 1974Thermoelasticity, pesudoelasticity and the memory effects associated with martensitic transformations (part 2 the macroscopic mechanical behaviour)," Journal of Materials Science, 9915361544

20. Liang Chen.1990Constitutive modeling of shape memory allcys," PhD Thesis, VPI, Blacksburg, VA.

21. Dimitris. C. Lagoudas, Iradj. G. Tadjbakhsh, 1992Active flexible rods with embedded SMA fibers," Smart Mater. Struct., 12162167
22. Lau Kin-tak, Zhou Limin, & Tao Xiaoming.2002Control of natural frequencies of a clamped-clamped composite beam with embedded shape memory alloy wires," Composites Structures, 5813947
23. Arup. K. Maji, Ihosvany. Negret, 1998Smart prestressing with shape-memory alloy," Journal of Engineering Mechanics, 1241011211128
24. Matsuzaki Yuji.1997Smart strutures research in Japan," Smart Mater. Struct., 64110
25. K. Otsuka, C. M. Wayman, 1998Shape Memory Materials," Cambridge University Press, Cambridge, UK; New York, USA.
26. Gangbing. Song, Brian. Kelly, Brij. N. Agrawal, 2000Active position control of a shape memory alloy wire actuated composite beam," Smart Mater. Struct., 95711716
27. A. V. Srinivasan, D. Mcfarland, Michael, 2001Smart Structures: Analysis and Design," Cambridge University Press, Cambridge, UK; New York, USA.
28. Sun Qingping, & Hwang Keh Chin.1993aMicromechanics modeling for the constitutive behavior of polycrystalline shape memory alloy- I. Derivation of general relations," J. Mech. Phys. Solids, 411117
29. Sun Qingping, & Hwang Keh Chin.1993bMicromechanics modeling for the constitutive behavior of polycrystalline shape memory alloy- II. Study of the individual phenomena," J. Mech. Phys. Solids, 4111933
30. K. Tanaka, 1986A thermomechanical sketch of shape memory effect: one-dimensional tensile behavior," Res Mechanica, 18251
31. K. Tanaka, S. Nagaki, 1982A thermomechanical description of material with internal variable in the process of phase transitions," Ingenieur-Archiv, 515287299
32. H. Warlimont, L. Delaey, R. V. Krishnan, H. Tas, 1974Thermoelasticity, pesudoelasticity and the memory effects associated with martensitic transformations (part 3 thermodynamics and kinetics)," Journal of Materials Science, 9915451555

33. Zhu Jiujiang, Liang Naigang, Huang Weimin, & Liu Zhihong. 2002A thermodynamic constitutive model for stress induced phase transformation in shape memory alloys," International Journal of Solids and Structures, 393741763

Chapter 3

Research on the Mechanical Properties of a Glass Fiber Reinforced Polymer-Steel Combined Truss Structure

Pengfei Liu[1], Qilin Zhao,[1] Fei Li,[2] Jinchun Liu,[3] and Haosen Chen[1]

[1]Institute of Field Engineering, PLA University of Science & Technology, Nanjing, Jiangsu 210007, China

[2]Department of Engineering Management, PLA Institute of Logistics Engineering, Chongqing 401331, China

[3]Institute of Defense Engineering, PLA University of Science & Technology, Nanjing, Jiangsu 210007, China

ABSTRACT

An assembled plane truss structure used for vehicle loading is designed and manufactured. In the truss, the glass fiber reinforced polymer

(GFRP) tube and the steel joint are connected by a new technology featuring a pretightened tooth connection. The detailed description for the rod and node design is introduced in this paper, and a typical truss panel is fabricated. Under natural conditions, the short-term load test and long-term mechanical performance test for one year are performed to analyze its performance and conduct a comparative analysis for a reasonable FEM model. The study shows that the design and fabrication for the node of an assembled truss panel are convenient, safe, and reliable; because of the creep control design of the rods, not only does the short-term structural stiffness meet the design requirement but also the long-term creep deformation tends towards stability. In addition, no significant change is found in the elastic modules, so this structure can be applied in actual engineering. Although the safety factor for the strength of the composite rods is very large, it has a lightweight advantage over the steel truss for the low density of GFRP. In the FEM model, simplifying the node as a hinge connection relatively conforms to the actual status.

INTRODUCTION

GFRP has excellent performance characteristics, such as lightness, high strength, flexibility in design, good antishock performance, and fatigue and corrosion resistance, as well as good protection and concealment function. Moreover, the cost is lower than that of carbon fiber reinforced polymer or basalt fiber reinforced polymer, which is more economically suitable for civil engineering. As a result, GFRP has been applied widely in civil engineering [1–8].

However, GFRP has some obvious defects: a lower elastic modulus and significant creep property. For instance, the tension elastic modulus of pultruded E-glass fiber/913 composite is 42 GPa at room temperature, which is 1/5 of steel's elastic modulus [9]. Meanwhile, the significant creep property alters the endurance performance of the structure and influences the long-term reliability of the structure [10]. Lifshitz and Rotem [11] find that, for a unidirectional glass fiber/polyester composite with a fiber volume fraction of 60%, creep failure occurs over 10^5 min for under 50% of the initial tensile strength at room temperature. Experiments show that creep is severe when the principal stress direction presents at 45° with longitudinal and latitudinal directions, and the total tensile creep strain of 100 h is about three times the initial strain, which is under 30% of the static strength, and the total creep deflection of 720 h is approximately 5.5 times the initial deflection. Under a stress higher than 50% of the static strength, bending creep failure occurs successively for the samples of 307 glass/polyester composites [12]. Additionally, the connection efficiency of frequently used connection technologies for a composites-adhesive connection, mechanical connection, and bolt-adhesive connection is not generally high, and there is a significant aging problem for the adhesive connection, which restricts the application of GFRPs in civil engineering as the main load-bearing structure. Therefore, the truss structure of GFRP is usually used for pedestrian bridge and roof structures that have a lower bearing capacity and a smaller span. For example, the bolted connection is adopted for a GFRP truss roof structure that is 19 m in length and 4 m in width to bear loads of wind and snow. For a GFRP truss bridge with spans of 12.5 m + 12.5 m and a 16.5 kN self-weight for each span in Switzerland, one span is connected by bolted joints, and the other span is connected by adhesive joints, which bears the pedestrian load [13]. Even if individual organizations developed the GFRP truss highway bridge, the structure is relatively heavier due to the limits of connection technology. Kostopoulos et al. [14] designed and manufactured a GFRP truss bridge, in which the composite members are held together by adhesive bonding and mechanical fastening joints. The bridge has a span of 11.6 m in length and 4 m in width and represents a single traffic lane. The bridge is Class 30 according to DIN 1072 (i.e., load bearing capacity of 300 kN) and weighs approximately 13.5 T, namely, 11.64 kN/m.

In this paper, considering the applications of and the development represented by existing GFRP truss structures in civil engineering as well as the demand for lightweight, modular bridges, for emergency purposes, a GFRP-steel combined truss structure is proposed. It is a vehicular bridge and is suitable for air-transport and manual erection by unskilled workers. It was designed as a modular bridge with a span length of 12 m and consists of 8 truss structural units. To achieve a low weight, GFRP was used for the chord, diagonal, and vertical members based on the consideration of the stiffness requirements of the whole structure. Moreover, a novel connector with a high connection efficiency, known as a pretightened tooth connection (PTTC) structure [15, 16], was employed for the connection of the GFRP material tubes.

Based on the proposed design, the weight of the truss panel is a mere 140 kg (40% lighter than the corresponding steel truss), and the bearing capacity of the truss bridge is comparable with the steel truss bridge of the same specifications.

The primary objective of this study was to verify the viability of the proposed GFRP-steel combined truss bridge. A description of the truss structure, including the conceptual design and fabrication is first provided. To investigate the structural performance, a single-layer and single-row assembled simply supported truss bridge with a length of 12 m was erected. The flexural properties of this specimen over both a short period of time and a long period of time were tested according to the principle of midspan moment equivalence. In addition, the corresponding numerical analysis based on two connection types was performed for a reasonable FEM model.

CONCEPTUAL DESIGN AND FABRICATION FOR A NEW COMPOSITE TRUSS

Conceptual Design for a Composite Truss Structure Type

For convenience in conducting the static load contrast experiment, the external dimension of a typical truss panel is in conformity with the external dimension of the "321" fabricated highway steel bridge truss panel (Figure 1), namely, a length of 3000 mm and a width of 1500 mm. However, for the weight reduction of all nodes, a cross-type truss of the fabricated highway steel bridge is not used for the GFRP-steel combined plane truss, but rather a parallel chord truss with fewer nodes is used. Members of the combined truss are made from a pultruded unidirectional glass fiber reinforced resin matrix composite and the joints are made from 16 Mn steel (Figure 2). They form an integrity that can transfer a load by pretightened tooth connection technology. Composite members of the truss are connected mutually at different type nodes, N-1, N-2, N-3, and N-4, through welding or bolt of steel joints (Figure 3).

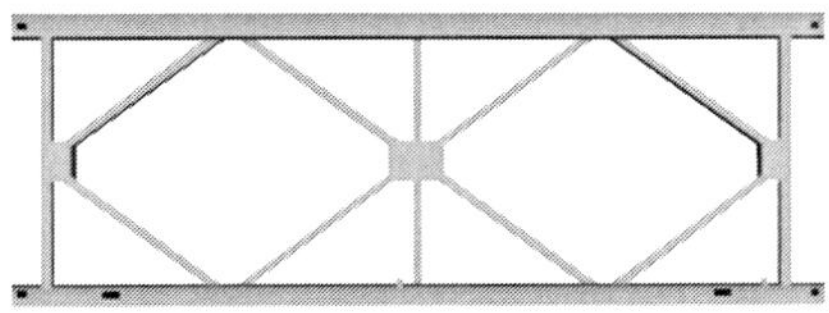

Figure 1: The "321" fabricated highway steel bridge truss panel.

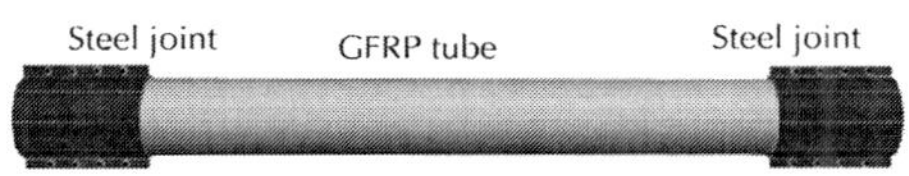

Figure 2: The integrity of the GFRP tube and steel joint.

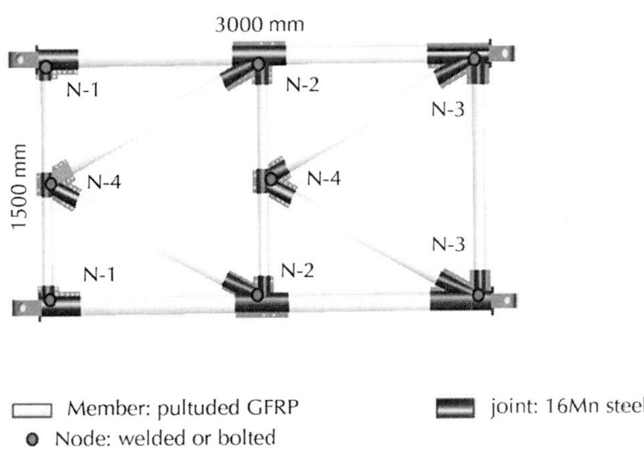

Figure 3: Structural design of the GFRP-steel combined truss.

Conceptual Design for Members

Common composite pultruded profiles are available in H-shaped, channel, and tubular composite materials. In the design, composite members are connected using pretightened tooth connection technology, and the cross-section of the member is shaped such that it is easy to connect the joints. Thus, tubular composite materials are used. Finally, pultruded glass fiber reinforced isophthalic unsaturated polyester resin composite tubes with a fiber mass fraction that is 61% manufactured by the Nanjing Jinjiuding Composite Company are chosen. Their density is 1.667 g/cm^3. Mechanical parameters provided by the manufacturer are shown in Table 1.

Table 1: Mechanical properties of glass fiber reinforced isophthalic unsaturated polyester resin composite tubes at 23°C and 50% relative humidity

Elastic modulus (GPa)			Shear modulus (GPa)			Strength (MPa)		
E_1	E_2	E_3	G_{12}	G_{13}	G_{23}	σ_t	σ_c	τ_{12}
25	3	3	1.2	1.2	9.6	1324	330	35

"1" indicates the fiber direction; "2" and "3" indicate directions that are perpendicular to the fiber. Direction 3 is orthogonal with direction 2. "t" indicates tension. "c" indicates compression.

The geometric size of the member depends on the forced state under actual working conditions. The GFRP-steel combined truss is guaranteed to bear the design load, which is not less than the bearing capacity of an existing "321" fabricated highway steel bridge. Namely, for a 12-meter long, single-row monolayer design, the wheel load is 20 tons in the midspan [17]. The internal force value of this span and load can be easily acquired from the structural mechanism model, as shown in Table 2. The internal force value of the upper and lower chords of the truss is the same, both of which are 200 kN. The upper chord is compressed, and the lower chord is tensed. However, it is considered that the truss can be used for both upper and lower chords, and the compressive strength of composites is obviously less than the tensile strength, so they are all designed as compressive members. To reduce the manufacturing cost and avoid remanufacturing the pultrusion mold, up and down composite tubes with a 104 mm external diameter and 8 mm thickness are selected. For the same reason, composite tubes with a 76 mm external diameter and 8 mm thickness are chosen to serve as the diagonal members and vertical members. Then, with reference to compressive strength, the cross-sectional area of the member, length of the member and elastic modulus of composites, different compressive strengths, and buckling loads for members, as well as safety factors, are determined; see Table 2 for details. Due to the dispersion of GFRP, the minimum safety factor for an actual application is 2 [17]. During the primary design, all of the safety factors of the members are more than 2 [18], while the strength and stability of the members meet the requirement.

Table 2: Size and performance of the composite truss members

Location	*L* **(mm) × number**	ED (mm)	*T* **(mm)**	IFE	DF	IFSF	CFB	SF
Chord	2800 × 2	104	8	200	250	796	718	2.9
Vertical	1292 × 3	76	8	25	37.5	564	302	8
Diagonal	1500 × 4	76	8	56	85	564	224	2.6

L: length; ED: external diameter; *T*: thickness; IFE: internal force under a 20 t external load; DF: designed internal force under external load and dead weight; IFSF: internal force for strength failure; CFB: critical internal force for elastic bulking; SF: safety factor.

Conceptual Design for Joints

A pretightened tooth connection processes teeth on the composite and transfers the axial force via the shear capability of tooth root, and the interlaminar compressive stress formed by imposing the pretightened force improves the interlaminar shear strength so that the composite can transfer larger load in joints [16] (Figure 4).

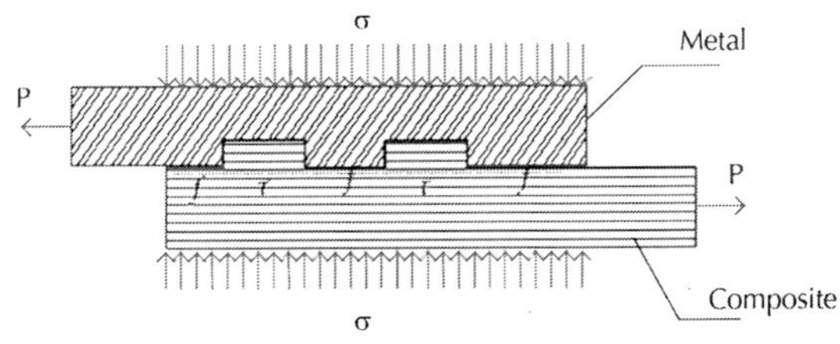

Figure 4: The mechanism diagram of a pretightened tooth connection for composites.

The circular tube joint of the composite made by pretightened tooth connection technology usually consists of external metallic sleeves with lugs, middle composite tubes, and internal metallic tubes. To fabricate such a joint, a metal tube that has an external diameter that is identical to the inner diameter of a composite tube that has greater stiffness is placed in the composite tube, and, then, the matching teeth are machined on the outer face of the composite tube and the inner face of the metal sleeve. In the last step, the metal tube is assembled along with the composite tube, and a great radial pressure stress is applied on the composite tube as high-strength bolts (Figure 5). The bearing capacity of this connection mode is related not only to the material properties but also to the pretightening force, tooth depth, and tooth length, and tooth quantity.

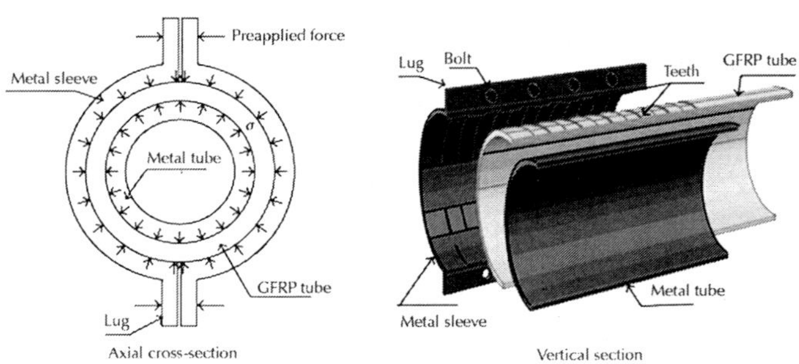

Figure 5: Schematic diagram of the joint.

The design for the member joints of the GFRP-steel combined truss structure mainly involves determining the geometry of the joint's steel sleeve, the tooth quantity, the tooth depth, and the length and value of the preload, based on the requirement for the bearing capacity. The joints can be divided into four categories depending on their locations and stress state (Figure 6): (1) chord end joint C-1: these types of joints are located on the ends of the chord member, are connected to the male or female heads, and serve as the major power transmission parts of the truss structure so they require a large number of teeth and increased preloads; (2) central joints of chord member C-2: these joints connect the vertical members with the diagonals and eliminate the need to pass large loads, thus requiring less teeth and lower preloads; (3) joints at the ends of the diagonal or vertical member C-3: these joints are located on the ends of the diagonals or the vertical members and only require passing the loads imposed on the diagonal or vertical members. As these are not heavy loads, they require less teeth and lower preloads; (4) joints in the middle of vertical member C-4: these nodes mainly work to connect the diagonals without passing heavy loads, thus requiring less teeth and preloads.

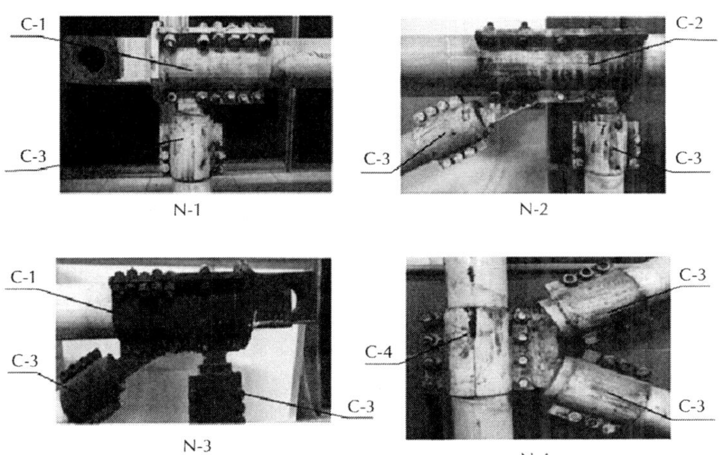

Figure 6: Classification of the joints in the GFRP-steel combined truss.

The design principles of the joints are as follows: (1) joints can transfer loads of no less than the design loads of the members. (2) Steel is used as little as possible at the joints for the weight reduction of the truss. Therefore, for end joints of the chord members of C-1 that have greater preloads and the joints of C-2 with a larger diameter, the outer steel sleeve needs to be divided into three sections with three pairs of lugs, while the other divided outer steel sleeves require two pairs of lugs. (3) The impact of the welding temperature on the composite tubes is taken into account when joints are fabricated. Based on these principles, in combination with the design method of the composite pretightened tooth connections and adopting safety factors larger than 2, the optimal depth and length of the teeth, the geometry of the joints, and the carrying capacity can be obtained, as shown in Table 3. The inner metal tube and outer steel sleeve are made of 16 Mn steel. There are two types of external diameters of the inner steel tubes, 88 mm and 60 mm, with an 8 mm wall thickness and a length matching the outer sleeves.

Table 3: Size and bearing capacity of the joints

Type	NT	DT (mm)	WT (mm)	*P* (kN)	TT (mm)	TS (mm)	PN	CC (kN)	DF (kN)
C-1	5	2	12	60	8	8	3	600	250
C-2	3	2	12	30	8	8	3	300	85
C-3	3	2	12	30	8	8	2	300	37.5
C-4	3	2	12	30	8	8	2	300	85

NT: number of teeth; DT: depth of tooth; WT: width of tooth; *P*: preload; TT: thickness of the inner steel tube; TS: thickness of the outer sleeve; PN: pair number of lugs; CC: carrying capacity; DF: designed internal force.

Design and Assembly for the Truss Joint

The pretightened tooth connection joints not only can be used for unidirectional extension or multidirectional connections of composite members but also can be easily made into single- or double-lug joints between GFRP-steel combined truss panels and spherical joints. Therefore, they can be widely used in truss bridges, space trusses, and prestressed bridge structures. The joint is designed according to the principle of reasonable construction, reliable connection, and convenient manufacture so that the existing mature weld technology of steel or bolted connections can be used.

The rod is simplified to have an axial tension and compression status. While designing the joint by the welding method, the C-3 joint is connected with other types of joints to form different nodes. First, a short steel column with a diameter that is equivalent to the inner diameter of the external sleeve is inserted and an all-around fillet welded at the end of the C-3-type joint. The top surface of this short column is used for welding with other joints. The width of the weld leg $h_f = 8$ mm, length of weld $l_w = 3.14 \times 76$ mm $= 238.64$ mm, strength of fillet weld $f_f^w = 165$ MPa, and the bearing capacity of fillet weld $[N] = 0.7h_f l_w f_f^w = 220.5$ kN is more than the maximum value of 85 kN of the design internal forces for diagonal rods and vertical rods and meets requirement. Second, through a K-type butt weld, the weld at the C-3 diagonal joint, and the lug of chord member joint C-1, the weld at the C-3 diagonal joint and the lug of vertical member C-4, the weld at the C-3 diagonal joint and the lug of chord member joint C-1, and the C-3 joints of the vertical member and diagonal member and the lug of chord member joint C-2 form nodes N-1, N-2, N-3, and N-4 (Figure 7). The butt weld bearing capacity $[N]$ is calculated as the most conservative three-level weld joint in tension [19], and the safety factors are uniformly greater than 2 (Table 4).

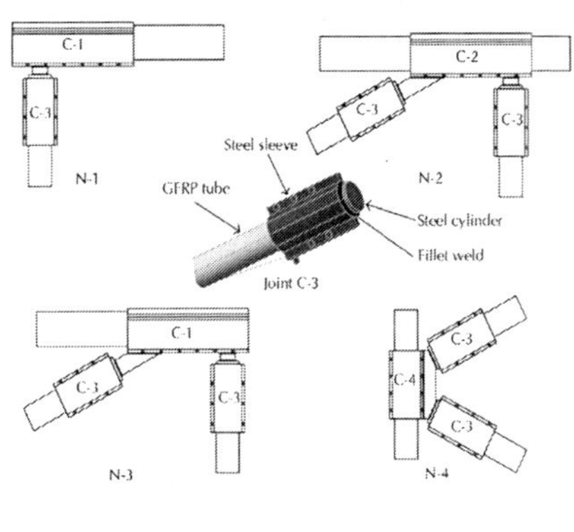

Figure 7: Different types of nodes welded by joint C-3 and other joints.

Table 4: Size and bearing capacity of the butt weld

Group	t (mm)	l_w (mm)	f_t^w (MPa)	[N] (kN)	DF (kN)	SF
Diagonals C-3 and C-2 or C-1	16	70	250	280	85	3.3
Diagonals C-3 and C-4	16	45	250	180	85	2.1
Vertical members C-3 and C-2 or C-1	16	40	250	160	37.5	6.7

t: weld thickness; l_w: length of weld leg; f_t^w: tensile strength of butt weld; [N]: bearing capacity of fillet weld; DF: designed internal force; SF: safety factor.

To the best of our knowledge, this is the first attempt to fabricate this type of truss structure, so there are no norms and standards in this respect. Therefore, the reference is made according to the welding procedures that are used for the existing assembled highway steel truss bridges and composite connection technologies, and these assembly and welding methods are proposed for the structure. This method can effectively reduce the stress associated with the assembly and ensure that all members can be accurately assembled. The method is divided into three steps.

Step 1

Steel joints are prepared, which mainly involves the processing of the inner steel tube, outer steel sleeves, and male and female heads at the ends. To prevent the composite tubes from being damaged by high temperatures while the joints are welded, the steel connection is somewhat longer than what is required for the joints to pass forces (Figure 8(a)). The truss panels are connected by male and female heads that are made specifically for this purpose.

Figure 8: Assembly process of the truss.

Step 2

Teeth are machined at the corresponding locations on the composite tubes and assembled with the inner steel tubes and the outer steel sleeve of joints (Figure 8(b)). To ensure assembling accuracy, the threads must be correctly positioned with accurate lengths and depths and perfectly matched with the teeth on the outer steel sleeve. While composite tubes are assembled with steel joints, all of the lugs used for welding must be in the same plane.

Step 3

Members are assembled. The assembly of members involves welding the steel joints at the ends and in the middle of the members. In the assembly, all of the welded members of the truss panel are in the same plane. To prevent the welding temperature from impacting the composite tubes, cooling measures, such as cold compression using a wet towel, are taken during welding. Chord members are installed first (Figure 8(c)). The specially made male and female heads are connected with the existing assembly highway steel bridge truss on both sides via single-pin connections, before the steel sleeves of the composite chord members are welded with the male and female heads. Then, vertical members are installed (Figure 8(d)). The steel sleeves on the ends of the vertical members are welded to the lugs of the chord members. The center line of the vertical member must coincide with the center of the lug of the chord member. Meanwhile, the lugs of the three vertical members remain in the same plane to ensure that the diagonals are in the same plane while they are welded (Figure 8(e)). Finally, the diagonals are installed (Figure 8(f)). The steel sleeves on the ends of the diagonals are welded to the lugs of the steel sleeves in the middle of the vertical members. The center line of the diagonal must coincide with the lug of the vertical member.

The assembled composite truss panel weighs 160 kg, which is 40% lighter than the 270 kg steel truss panel of "321" assembled highway steel bridges.

MECHANICAL PERFORMANCE TEST AND ANALYSIS OF A GFRP-STEEL COMBINED TRUSS STRUCTURE

Structure Test Scheme

Short-Term Loading and Test Scheme

To verify the reliability of the combined truss panel, a single-layer and single-row assembled simply supported highway steel bridge with a length of 12 m is erected. Therein, Side-A consists of four steel truss panels and Side-B consists of three steel truss panels and one combined truss panel in the midspan. According to the principle of midspan moment equivalence, the mode that the steel members and steel tanks are stockpiled on the platform of deck slabs and the water is injected into the box shall be used to simulate a 20 t wheel load. The load is transferred to the main truss through four beams in the midspan (Figure 9).

Figure 9: Static live loading mode of the structure.

The deflection of the bridge and the stress of the key rods and joints are tested during loading. Dial indicators with manual readings are set in the location of the midspan, *L*/4, and support position of each side. Five test points are set at each side, totaling 10 test points on both sides. The numbering is from D1 to D10. The strain gages are set on the upper chord, lower chord, and diagonal members of the truss in midspan of Side-A. The numbering is from S1 to S4, and the internal force of steel rod is tested. The strain gages are set on the upper chord, lower chord, and vertical and diagonal members of the combined truss in the midspan of Side-B. Its numbering is from S5 to S11, and the state of composite rods shall be tested. In addition, the strain gages are set on joint C-1 at the end of the lower chord of the combined truss panel. The numbering is from S12 to S18, and the stress state is tested (Figure 10).

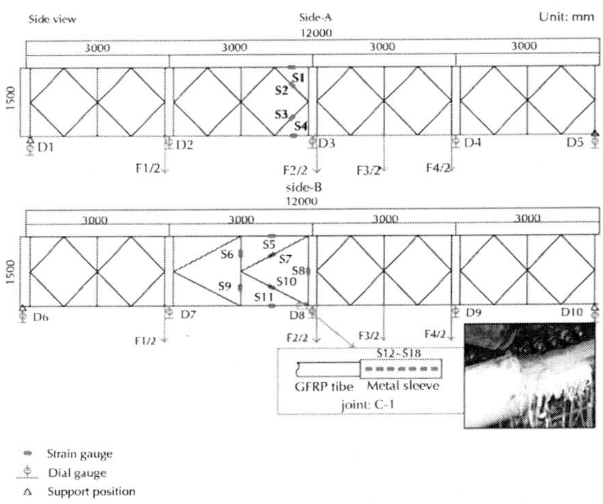

Figure 10: Static live loading position and sensor arrangement.

The equivalent bending moment starts at 0 and progressively increases to 600 kN·m under 6-step loadings. The bearing reactions distributed to the main truss by transverse beams during step loading is maintained for 2 mins after each step loading for the sufficient development of deformation, and structural deflections and stress are collected, see Table 5. The truss is monitored in real time during loading; in cases of the obvious bending of the composite rods, obvious looseness of the steel joints and rods, or notable broken soundness of the material, the loading is stopped immediately.

Table 5: Step loading in the experiment

EBM (kN·m)	120	240	300	360	480	600
F_1 (kN)	8.9	17.8	22.3	26.7	35.6	44.5
F_2 (kN)	20.7	41.5	51.9	62.2	83.0	103.7
F_3 (kN)	11.8	23.7	29.6	35.5	47.3	59.2
F_4 (kN)	11.8	23.7	29.6	35.5	47.3	59.2

EBM: equivalent bending moment

Test Scheme for Long-Term Loading

The long-term performance of the truss affects the safety of the truss structure. The long-term performance indexes consist of the reduction rate of the existing bearing capacity of the truss structure, deflection of the truss structure, and corrosion and damage degree of the truss structure. In view of the more obvious creep property of GFRP, long-term deflection monitoring and research on stiffness changes one year later are conducted for the truss structure.

The load for the long-term deflection test is the sum of the self-weight of the truss structure and the steel weight on the loading platform. The midspan equivalent moment is 300 kN·m; that is, the upper chord of composite truss bears 42 MPa, of which 13% of the stress is the ultimate strength and is lower than the limitation of 30% for long-term creep control stress [20, 21]. The long-term deflection change is tested with balance level.

One test point per 3 m is arranged under the lower chord of the truss on both sides and 5 are arranged on each side, totaling 10, the position of which is identical to that of the dial indicators in Figure 11. The survey has been performed continuously for over one year from March 2011 to April 2012, with a frequency of once per 15 days during the first three months, then once per month during the second three months, and, finally, once per half-year after six months. The temperatures were between −8°C and 39°C, with an average temperature of 16.4°C and an average relative humidity of 76%.

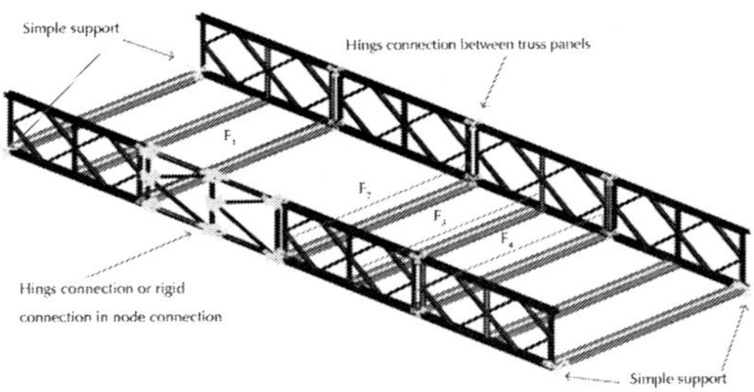

Figure 11: The finite element method model.

After one year, in April 2012, the truss is added with the step equivalent bending moment of static live loads 60 kN·m, 120 kN·m, and 240 kN·m according to the short-term performance test method from one year earlier. The total added load is 540 kN·m. The change of one side deflection for the GFRP-steel combined truss structure is tested during the loading; the test point distribution is shown in Figure 10.

Structural Mechanical Property Analysis under a Short-Term Load

FEM Model

The FEM analysis software ANSYS 12.1 developed by ANSYS, Inc., in Pennsylvania, USA, is used to simulate the stress and deformation under midspan equivalent bending moments of 120 kN·m, 240 kN·m, 300 kN·m, 360 kN·m, 480 kN·m, and 600 kN·m that are generated by loadings F_1, F_2, F_3, and F_4 on four beams in mi-span scattered across both sides in the experiment. In the case of two simple supported ends, the deflection, axial stress, and axial strain under a load with various levels can be obtained (Figure 11).

Regarding the steel truss structure, the Mohr formula is usually adopted for a theoretical deflection calculation to simplify the node to hinge connection, and only the axial tension and compression stress are considered. Due to large steel modulus, disregarding such secondary stress, as the shear stress in the Mohr Formula is feasible, but for the resin matrix composite with a low modulus, it is unknown whether the effect should be considered in the new truss structure. Therefore, BEAM188 element modeling is used in consideration of the above situation for increased accuracy, with values that are closer to the actual value. Please refer to Tables 11 and 22 for the parameters and cross-section forms of the composite truss. For the GFRP-steel combined truss structure, the performance in nodes between composite and steel is different. Because it is still unknown whether they can be simplified through a calculation, when building models, analogue experiments for both the hinge connection and rigid connection in the node connection are conducted to obtain the result that determines which one is more suitable.

Analysis of the Structural Stiffness Performance

The vertical stiffness of a truss is related to the cross-section, elastic modulus, and structural distribution, which is inversely proportional to the deflection. The elastic deflection generated by the static live load of the truss bridge is the main control value for verifying the vertical stiffness [22]. As a result, the calculated value and experimental value of deflection for this truss bridge are analyzed.

Under different step loadings, deflections on both sides of the truss structure that change in the longitudinal direction are indicated in Figures 12 and13. It turns out that, in either Side-A or Side-B, with the combined truss, in each phase of loading, the deformation of the truss is the same as well as the maximum deflection on both sides of truss structure in the midspan. Owing to the stiffness of a GFRP-steel combined truss structure being lower than that of a steel truss, the deflection on Side-B is distinctly larger than that on Side-A (Figure 14). Under operating conditions of maximum loading, the displacement on Side-B is 1.67 times that of Side-A, which proves that the stiffness of glass fiber structure is most probably the control factor of this structure.

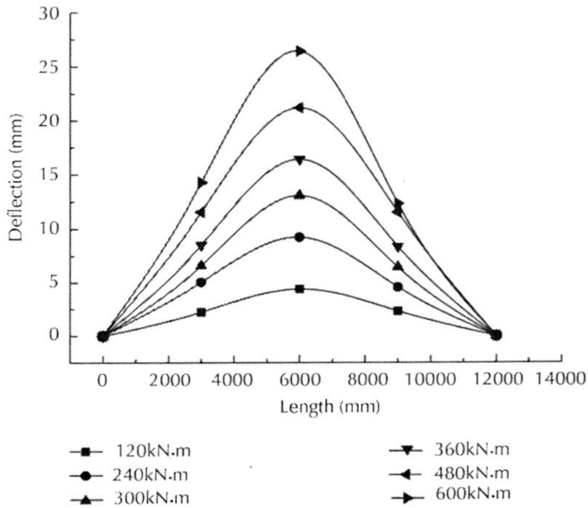

Figure 12: Deflection of Side-B with GFRP-steel combined truss.

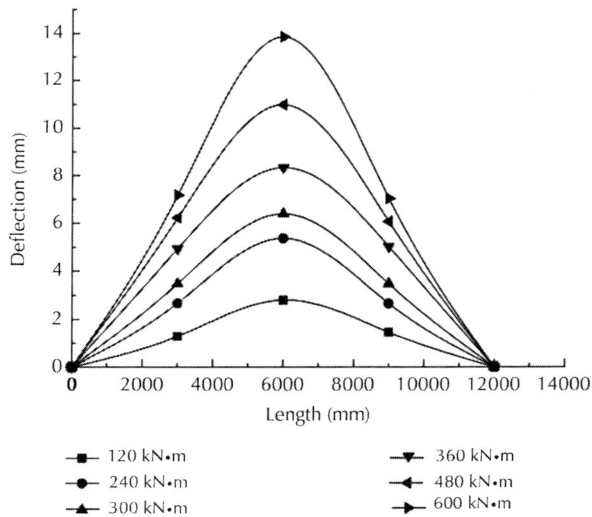

Figure 13: Deflection of Side-A.

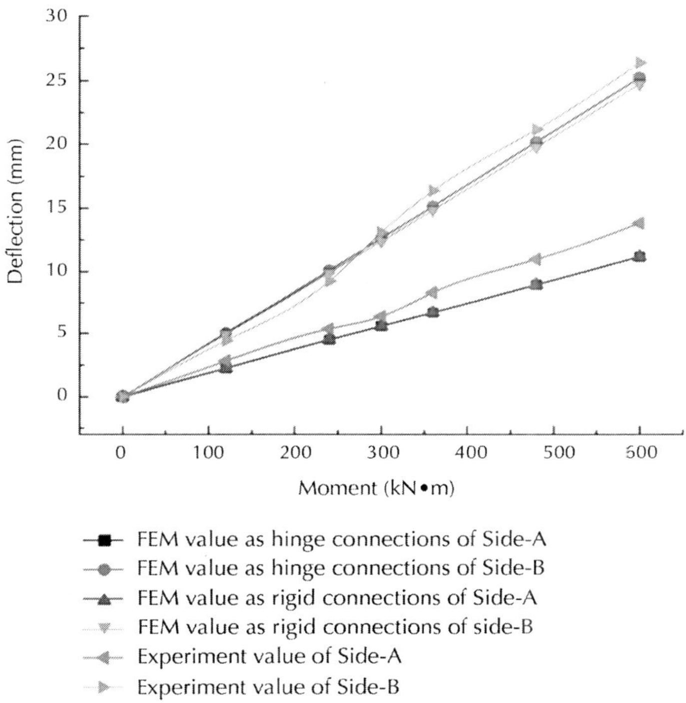

Figure 14: Finite element calculated values and the experimental values of deflection in the midspan for both sides.

Meanwhile, after comparing the maximum experimental value of deflection on both sides of the truss structure in the midspan with the FEM value, the difference is distinct but tends to be of a stable value. This is because the experimental value of deflection for fabricated highway steel bridges includes elastic deflection and inelastic deflection; nevertheless, the calculated FEM value is only the elastic deflection under a static live load. The elastic deflection is the elastic deformation produced under the effect of dead weight and the static live loading of the truss structure. The inelastic deflection is the inelastic deformation of the structure caused by relative displacement because of the design clearance between the pins and the holes of pin joints in the truss

[22]. Generally, due to the friction between the pin and the hole, the inelastic deflection will gradually develop to the maximum, with an increase in the static live load. Therefore, the difference between the experimental value and the theoretical value under the loading is the residual inelastic deflection.

From Figure 14, for Side-A, the difference in the calculated finite element values between the rigid connection and hinge connection, both of which are transferred from the internal nodes of the steel truss panel, is slim, whereas, the value for the hinge connection is larger than the rigid connection when transferred from the internal nodes of the combined truss panel in Side-B, which approaches the experimental value. Accordingly, converting the nodes inside the combined truss panel to the hinge connections is more practical and simplifying, as the hinged connection is suitable because the elastic modulus of the composite is lower, and when one rod is rotating, not only does the opposite joint follow suit easily but also the restrain from the other rods is also relatively slim.

To determine whether the deflection for the simply supported truss bridge with a 12 m span, which uses combined truss panels, meets the specified limitation under a 20 t wheel load, the finite element ANSYS software is utilized to build the finite element model, and a 200 kN static live load is imposed on midspan in this model. The maximum midspan deflection calculated is 66.003 mm. The design specifications on the simple truss steel bridge for emergency use provide that when the span $L < 50$ m, the vertical elastic deflection generated by the static live load is $f \leq L/150$. In this case, it is 80 mm. This indicates that the new composite truss bridge satisfies the stiffness requirements.

Analysis of the Structural Strength Performance

By step loading, the relationship between the equivalent bending moment in the midspan on both sides and strain of the rods is determined, as shown in Figures 15 and16.

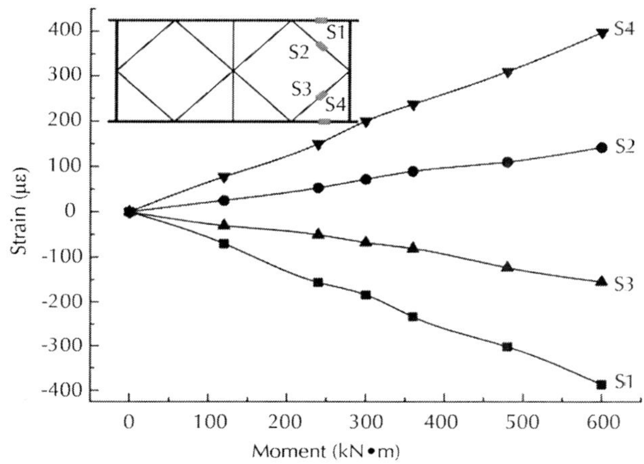

Figure 15: The relationship between the equivalent bending moments and experimental values of strain for Side-A.

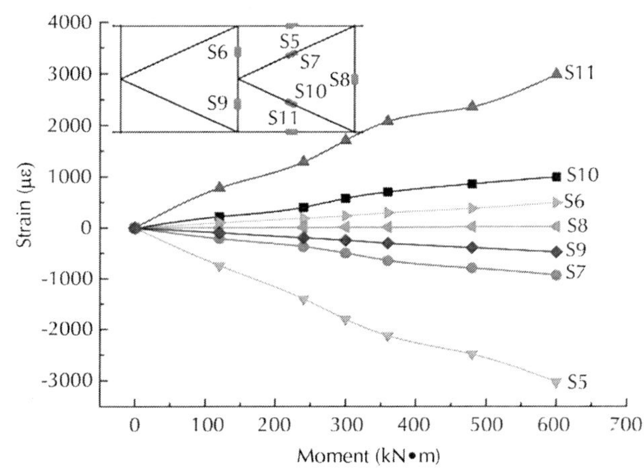

Figure 16: The relationship between the equivalent bending moments and experimental values of strain for Side-B.

From Figures 15 and16, in either the steel truss member or the composite rod, the overall bending moment-average longitudinal strain changes in a linear form. All of the strain values for the upper chords are negative, while all of the strain values for the lower chords are positive, but, basically, their absolute values are the same. The same is true for the diagonal rod and vertical rod. This is in agreement with the theoretical analysis. At the same time, by the combination of the elastic modulus with a cross-sectional area of the rod, the experimental internal force value and FEM value (Table 6) for combined truss rods can be obtained at a value under 600 kN·m. The FEM value is well matched with the experimental value; therefore, the model is accurate. The experimental value and FEM value are both less than the design value, while the rod strength meets the requirement and the rods are provided with large safety factors.

Table 6: Contrast experimental value with the FEM value of composite rods under 600 kN·m

Member	S5	S6	S7	S8	S9	S10	S11
IFD (kN)	200	25	56	25	25	56	200
EV (kN)	−182.4	19.8	−42.5	1.3	−19.9	42.8	180.3
FEM (kN)	−190.8	20.5	−43.9	1.4	−20.7	43.9	183.3
Deviation	4.6%	3.5%	3.3%	7.7%	4%	2.6%	1.7%

IFD: internal force value for 20 t designed loading; EV: experimental value; deviation: deviation = |(EV − FEM)/EV| × 100%.

The main concern of the experiment is that joint C-1 pulls on the lower chord without determining the relative sliding between the composite tube and metal sleeve. Refer to Figure 17 for the relationship between the average longitudinal strain and the bending moment of steel joint C-1 on the lower chord of GFRP-steel combined truss structure. When the lower chord is subjected to pull, the test point should be positive, in theory, but it appears that the positive and negative strains alternate in the joint, of which the negative strain is caused by the eccentric bending moment that is produced by the load on the internal teeth of steel sleeve. Regarding various strains in different test points and loads, the changing rule is more or less constant. The average longitudinal strain on the surface of the steel joint becomes larger linearly with an increase in the bending moment. Under the maximum bending moment of 600 kN·m, the maximum longitudinal strain is 87.5 μ, which is less than the yield strain 1668 μ of 16 Mn steel and without relative sliding. It proves that the pretightened tooth connection technology of the composite is reliable under the design load.

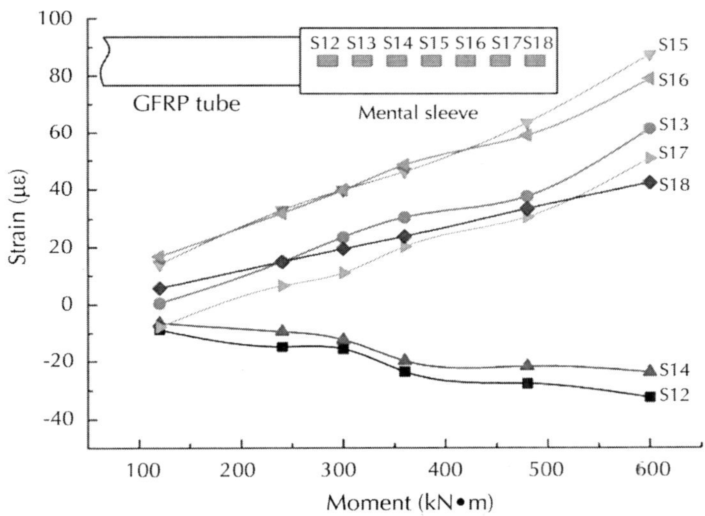

Figure 17: The relationship between the average longitudinal strain and bending moment of steel joint C-1 on the lower chord of a GFRP-steel combined truss.

Analysis of the Structural Mechanical Property under a Long-Term Load

The study of the mechanical property under long-term loading is based on the creep effect of the composite to the truss structure, namely, aiming at the long-term deflection and stiffness of the combined structure.

Figure 18 shows the relationship between the deflection increment of the midspan on both sides of the truss structure and time.

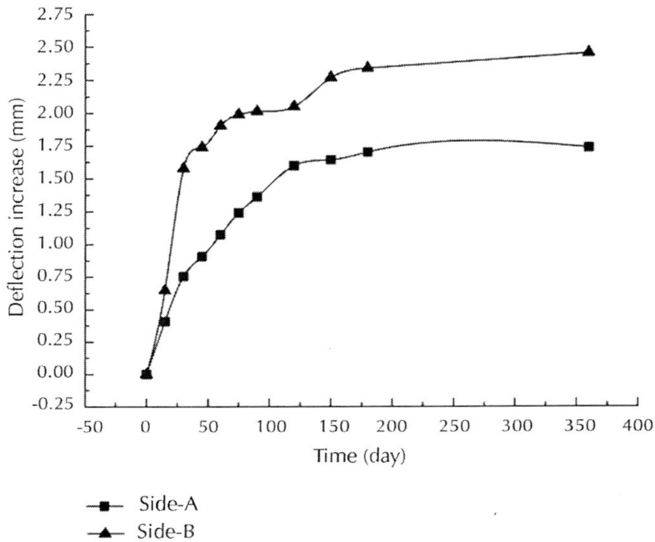

Figure 18: The relationship between the deflection increment of the midspan on both sides of the truss structure and time.

From Figure 18, the deflection of Side-A of the truss structure increases with time. After 360 days, the total change is 1.73 mm, accounting for 27% of the short-term experimental value of 6.381 mm under 300 kN·m, and the change rate for the deflection increment tends to be stable over time. The deflection on Side-B increases with time, and, after 360 days, the total change is 2.45 mm, accounting for 19% of the short-term experimental value of 13.064 mm under 300 kN·m; the change rate for the deflection increment tends to be stable over time. The deflection change rate of Side-B with the GFRP-steel combined truss is 42% larger than that of Side-A without the GFRP-steel combined truss, not exceeding 3 mm in total, which verifies that the long-term performance of the GFRP-steel combined truss is very stable.

Composite materials are easily affected by time, temperature, humidity, and other factors, which cause a change in mechanical properties. One important manifestation is the relationship of elastic deflection increments with bending moment increments. From Figure 19, the relationship curve between the midspan deflection and the bending moment from one year ago basically parallels the relationship curve between the midspan deflection and the bending moment from one year later. The change trend is consistent, namely, the overall stiffness of Side-B with the combined truss panel that remains unchanged shows that the long-term exposition property of the GFRP-steel composite truss structure is stable under natural conditions.

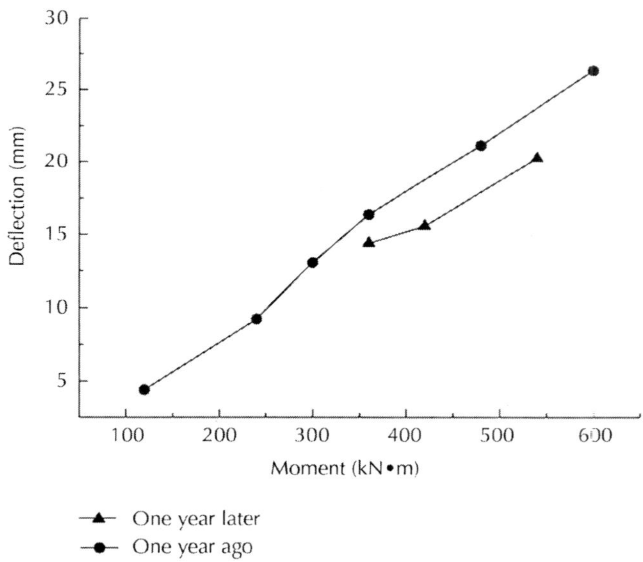

Figure 19: The relationship curve between midspan deflection and the bending moment for Side-B during one year.

The loading in the test increased stepwise up to 540 kN·m. As there is no relative sliding at the connecting part during loading, the pretightened tooth connection joint is reliable.

The analysis shows that the GFRP-steel combined truss has a stable performance and is suitable to use for temporary emergency bridges.

CONCLUSIONS

In this study, a GFRP-steel combined truss is put forward to make a typical truss panel. Furthermore, load tests and analyses are conducted in a 12 m span with a 20 t equivalent wheel load. The research shows the following.

i. The ends of composite rods are fabricated with steel joints by way of pretightened tooth connections. The mature weld technology is convenient for the design and manufacture of various nodes of the composite combined truss, and the experiment proves the reliability of the transmission of a load.

ii. The structural deflection of the designed combined structure with a 12 m span under a 20 T equivalent wheel load is 66.003 mm less than the standard specified *L*/150, namely, 80 mm. The short-term stiffness meets the requirement. While the long-term stress state of the GFRP rod is less than 30% of the design strength, the structural long-term creep deformation tends to be stable over 3-4 months and the long-term deformation is 1.19 times the previous elastic deformation.

iii. One year later, the actual test for the stiffness of the combined truss structure is conducted. Compared with the stiffness at the beginning of its erection, the stiffness of the truss structure does not decrease, which indicates that the elastic modulus of the glass fiber composite material does not decrease significantly under natural conditions and no relative sliding occurred at the pretightened tooth connecting joints; therefore, the structural stiffness is reliable.

iv. In the span and design load, the weight decreases by 40%, with the steel truss panel weighing 270 kg and the GFRP-steel combined truss panel weighing 160 kg, which proves that the combined truss has an advantage of being lightweight.

In the future, the fatigue performance of the combined structure will be investigated.

ACKNOWLEDGMENTS

This work was supported by the Major State Basic Research Development Program of China (973 Program, Grant no. 2012CB026202) and the National Natural Science Foundation of China (Grant no. 11372355).

CONFLICT OF INTERESTS

The authors declare that there is no conflict of interests regarding the publication of this paper.

REFERENCES

1. Gilby J. Pultrusion provides roof solution. *Reinforced Plastics*. 1998;42(6):48–52.
2. Gonilha JA, Correia JR, Branco FA. Creep response of GFRP-concrete hybrid structures: application to a footbridge prototype. *Composites B: Engineering*. 2013;53:193–206.
3. Gonilha JA, Correia JR, Branco FA. Dynamic response under pedestrian load of a GFRP-SFRSCC hybrid footbridge prototype: experimental tests and numerical simulation. *Composite Structures*.2013;95:453–463.
4. Peirick L. *Connection development and in-plane characteristics of innovative 3-D GFRP composite structural insulated panels for sustainable civil engineering infrastructure applications [M.S. thesis]*Houston, Tex, USA: University of Houston; 2011.
5. Santos Neto ABDS, La Rovere HL. Composite concrete/GFRP slabs for footbridge deck systems.*Composite Structures*. 2010;92(10):2554–2564.
6. Correia J, Cabral-Fonseca S, Branco F, Ferreira J, Eusébio MI, Rodrigues MP. Durability of glass fibre reinforced polyester (GFRP) pultruded profiles used in civil engineering applications. Proceedings of the 3rd International Conference on Composites in Construction; 2005; Lyon, France. pp. 11–13.

7. Jeong J, Lee Y, Park K, Hwang Y. Field and laboratory performance of a rectangular shaped glass fiber reinforced polymer deck. *Composite Structures*. 2007;81(4):622–628.
8. Mendes PJD, Barros JAO, Sena-Cruz JM, Taheri M. Development of a pedestrian bridge with GFRP profiles and fiber reinforced self-compacting concrete deck. *Composite Structures*. 2011;93(11):2969–2982.
9. Wang T. *Finite element analysis for simulate composite wing of Su-27 and higher-order theory [M.S. thesis]* Shenyang, China: Shenyang Aerospace University; 2012.
10. Yang T. *Viscoelastic Mechanics*. 1st edition. Wuchang, China: Huazhong University of Science and Technology Press; 1992.
11. Lifshitz JM, Rotem A. Time-dependent longitudinal strength of unidirectional fibrous composites. *Fibre Science and Technology*. 1970;3(1):1–20.
12. Shanghai FRP Research Laboratory. *FRP Structure Design*. 1st edition. Shanghai, China: China Building Industry Press; 1980.
13. Bai Y, Keller T, Vall EET. Dynamic behavior of an all-FRP pedestrian bridge. In: Smith ST, editor. Proceedings of the Asia-Pacific Conference on FRP in Structures; 2007; International Institute for FRP in Construction; pp. 1075–1080.
14. Kostopoulos V, Markopoulos YP, Vlachos DE, et al. Design and construction of a vehicular bridge made of glass/polyester pultruded box beams. *Plastics, Rubber and Composites*. 2005;34(4):201–207.
15. Li F. *Research on new type of composite connection and application in the truss [Ph.D. thesis]* Nanjing, China: PLA University of Science and Technology; 2012.
16. Ma Y, Zhao Q. Analysis of the bonded-bolted hybrid composite joints' carrying capacity. *Acta Materiae Compositae Sinica*. 2011;39(4):225–230.
17. Huang S, Liu M. *Multi-Purpose Manual of Assembled Highway Steel Bridge*. 1st edition. Beijing, China: People's Communications Press; 2001.
18. Sedlacek G, Trumpf H, Castrischer U. Development of a lightweight emergency bridge. *Structural Engineering International*. 2004;14(4):282–287.

19. Cao S. *The Principle of Engineering Structure Design*. 4th edition. Nanjing, China: Southeast University Press; 2002.
20. Scott DW, Zureick A. Compression creep of a pultruded e-glass/vinylester composite. *Composites Science and Technology*. 1998;58(8):1361–1369.
21. Batra S. *Creep rupture and life prediction of polymer composites [M.S. thesis]* Morgantown, WVa, USA: College of Engineering and Mineral Resources at West Virginia University; 2009.
22. Hu Y. *Military Bridges*. Beijing, China: PLA Publishing Press; 2001.

Chapter 4

Effect of Cavity Design on the Strength of Direct Posterior Composite Restorations: An Empirical and Fem Analysis

V. Susila Anand[1,2] C. Kavitha[1], and C. V. Subbarao[1]

[1]Conservative Dentistry and Endodontics, Saveetha Dental College, Saveetha University, Madras 600077, India

[2]Bioengineering Laboratory, Engineering Design Department, Indian Institute of Technology Madras, Madras 600036, India

ABSTRACT

The aim of the present study was to verify the hypothesis that cavity design does not affect the strength of direct composite restorations as do material properties. Finite element modeling (FEM) and empirical testing were done for two cavity designs: a box shape (cube) and a concave shape (U). Two microhybrid composites were used to prepare the samples with the help of split stainless steel moulds. Compressive strength was tested. The results were statistically analyzed. Both FEA and empirical testing were complementary to each other in that the concave shape showed a significantly higher strength than box. Material properties affected the values only when box shape was used. The null hypothesis is thus rejected, and it is concluded that design significantly affects the strength of direct composite restorations.

INTRODUCTION

Composite restorations have revolutionized restorative dentistry due to their conservative nature, adhesive bonding and patient appeal. The adhesive bonding, ability makes it unnecessary to remove tooth structure for retention, prevention, and convenience. Successful restorations can be done with less precise preparations.

With paradigm shift from retentive restorations to conservative restorations, there is an increasing emphasis on minimally invasive cavity preparations. There are many studies reporting the effect of cavity design on the fracture resistance of teeth restored with indirect composite restorations [1]. Other studies [2, 3] have evaluated the effect of cavity design on the marginal leakage of composite restorations. One of the earliest applications of engineering principles in cavity design, with the objective of minimizing stress concentration was by Bronner [4]. Further work was subsequently reported by Gabel [5], Brown [6], and Weiland [7]. The conclusions of these studies were that the cavity preparation should enhance the properties of the restorative material in such a way that those unfavorable are properly compensated for in terms of stresses; the cavity preparation has to allow the operator to work efficiently in such a way that a mechanically sound preparation is obtained [8].

Noonan [9] provided one of the first studies of cavity design utilizing two-dimensional photoelastic stress analysis. From this followed a great number of studies using the photoelastic technique [10–13] that investigated the effect of differing Class II cavity designs on the stress dissipation with the remaining tooth structure and the restorative material.Granath and Hiltscher[14] complemented his analog studies by strain gauge measurements when investigating the effect of the buccolingual shape, while Fisher and Caputo [15] investigated both extracoronal and intracoronal preparations. The finite element method was first introduced into the area of stress analysis of biological structures in 1972 by Brekelmans et al. [16]. But there is a dearth in studies on the effects of cavity design on the strength of direct composite restorations.

The aim of the present study is to evaluate the effect of cavity design on the strength of direct composite restorations. For this, we propose a null hypothesis that cavity design does not affect the strength of direct composite restorations as do the material properties. To validate this hypothesis, two cavity preparation designs, a conventional box and a new minimally invasive concave shaped cavity with 4° taper, were studied with two composite restorative materials, Restofill and Esthet-X.

MATERIALS AND METHODS

Two split stainless steel (s.s) moulds with suitable plungers were fabricated using prototype designing with Pro E software. The moulds were made of two different shapes, box (cube—group I; Figure 1) and concave shape (U design with 4° taper—group II; Figure 2). The internal line and point angles were rounded and the dimensions were 10 mm × 10 mm × 10 mm (height × width × length). The internal radius of "U" in concave shape was 5 mm. The inner surfaces of the moulds were smooth and well-polished and millimeter markings were inscribed on one wall to facilitate incremental composite curing.

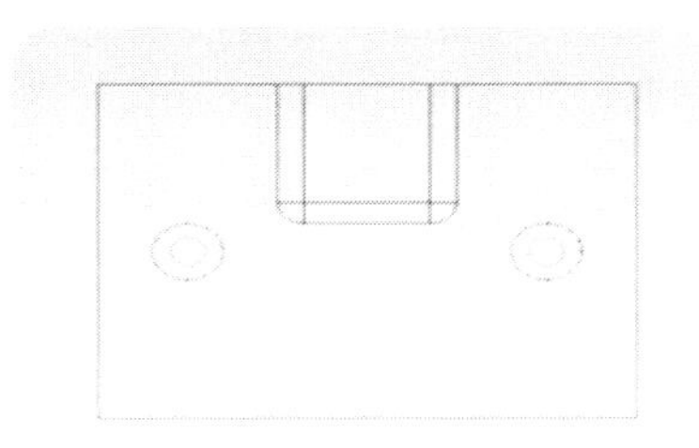

Figure 1: 2D image of s.s mould prototype—box design.

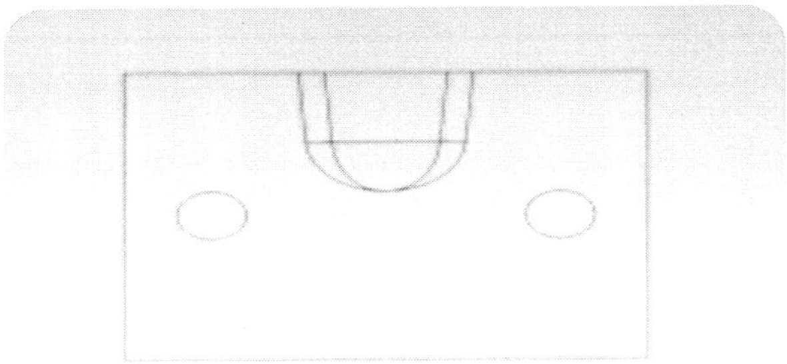

Figure 2: 2D image of s.s mould prototype—concave design.

Two microhybrid composites, Restofill (subgroup A) and Esthet-X (subgroup B) were used in the study. The composition of the materials is given in Table 1. Twenty samples were prepared for each design, 10 for each material. The material was condensed into the moulds using Teflon-coated composite condensers and pressure was applied to remove voids with the help of the built in s.s plunger of the mould. Every 2 mm increment of composite was cured with QTH curing light with an intensity of 400 mW/cm^2 for 40 s. The four corners of the moulds were focused for 10 s each with the curing tip to ensure adequate curing. The final increment was covered with a mylar strip during curing.

Table 1: Composition of composite materials used.

Esthet-X (DENTSPLY)	Restofill (Anabond Stedman)
Dimethacrylate resins	Bis-GMA
Urethane modified Bis-GMA	TEGDMA
Barium boroaluminosilicate glass	Barium boroaluminosilicate glass
Barium fluoroaluminosilicate glass	

The split mould design ensured that the specimens were retrieved easily without stress.

Empirical Testing

The composite samples were loaded under compression in a universal testing machine (Instron) at a cross-head speed of 1 mm/min. The load at fracture was noted and the compressive strength calculated.

FEM

Abaqus 6.7 (finite element tool) was used to create the 3-dimensional models of the above moulds using high-resolution images. A cube and a U-shaped model with10 × 10 × 10 mm dimensions were designed. (Figures 3 and 4) The internal radius of "U" was 5 mm. The models had the following physical characteristics: elastic modulus of 20 GPa, Poisson's ratio ( ) of 0.3, and density of 700 kg/m^3. Dynamic explicit procedure type was chosen for finite element analysis with encastre boundary conditions. A mesh size of 1 mm and an element size of 0.1 mm were used for the study. The von Mises stresses were calculated at 1500 N load with the Abaqus 6.7 EF1 software.

Figure 3: Cube mesh model.

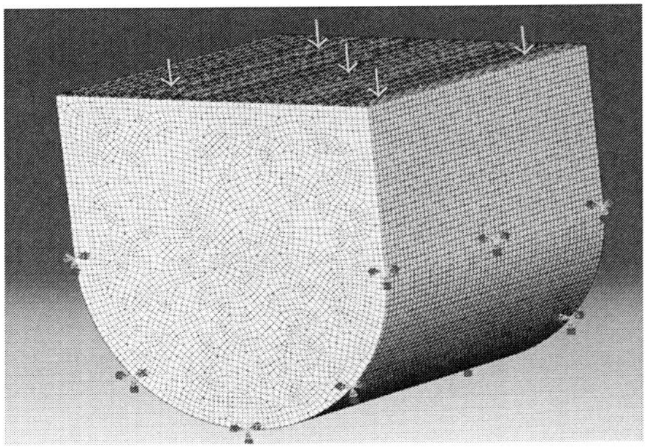

Figure 4: U-shaped mesh model.

Statistical Analysis

t-test was done to analyze the statistical significance of the results. Levene's test of Equality of variances and*t*-test for equality of means was done for intra- and intergroup comparisons of strength values.

RESULTS

Empirical Testing

The compressive strength was found to be 33 MPa (group I A box design, Restofill), 21.25 MPa (group I B Box design, Esthet-X), 49 MPa (group II A conservative design Restofill), and 49.49 MPa (group II B concave design Esthet-X; Table 2).

Table 2: Results of FEM

Group	Shape	Strength (MPa)	% improvement
I	Box	25.6	59
II	Concave	66	

FEM

The von Mises stress for group I (cube mesh) was found to be 25.6 MPa (Figure 5) and for group II (U-shaped model) 66 MPa (Figure 6) at a load of 1500 N. There was 59% improvement in the strength when the concave design was used (Table 3).

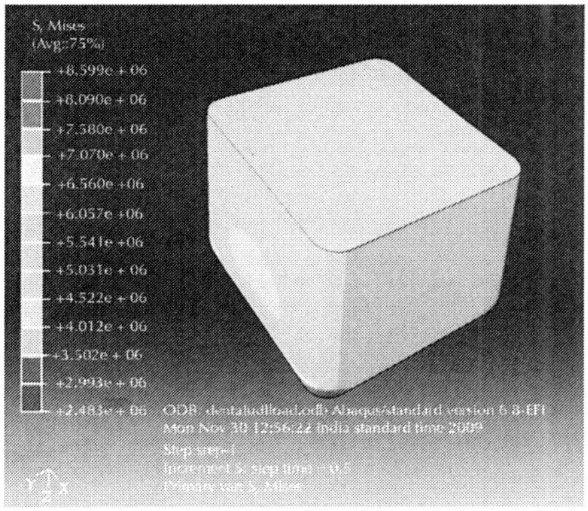

Figure 5: Finite element analysis (FEA) of cube model—box-shaped cavity design.

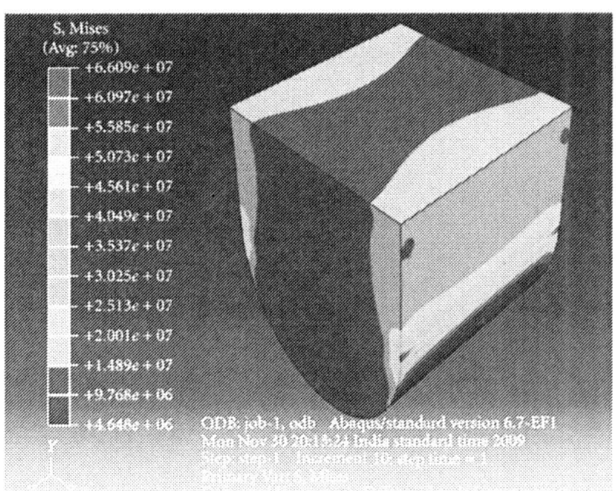

Figure 6: FEA of U-shaped model—concave cavity design.

Table 3: Strength of experimental groups.

Group	**Sub-group**	*n*	**Mean Strength (MPa)**	**Standard deviation**	**% improvement**
I	A	10	33	2.25918	33
	A	10	49	2.34434	
II	B	10	21.25	2.79466	57
	B	10	49.49	3.48038	

Empirical test showed statistically significantly greater strength for group II than group I ($P = 0.001$; Table 2) The results of FEM also showed greater values for Group II than I (Table 3). Among subgroups, A had significantly greater strength (Table 2) than B when box cavity (group I) was used ($P = 0.001$). When concave-shaped cavity was used, there was no difference in the strength values between the 2 materials ($P \geq 0.05$). When the concave design was used subgroup A showed 33% and subgroup B showed 57% improvement in the strength.

DISCUSSION

The present study proposed a null hypothesis that cavity design does not play a significant role on the strength of direct composite restorations. Based on this, 2 cavity designs were tested. One of the first studies on the effect of cavity preparation on the strength of teeth was by Vale [17]. Effect of width of isthmus preparation was studied and it was found that there was no difference in the ultimate fracture strength of the sound teeth with the 1/4 intercuspal distance cavity preparation. However, when the isthmus width was increased from 1/4 to 1/3 intercuspal distance, there was a significant weakening of the prepared tooth [8].

A study on two-stage shape optimization process for cavity preparation found that the stress level at the tooth-restoration interface in the optimized design was reduced significantly compared with the conventional design, irrespective of the bonding condition [18]. In the present study two cavity designs were evaluated for their effect on the strength of two different composite restorative materials. It was found that the conservative design (concave shape) had better strength (twofold on an average) than conventional design irrespective of the type of material. The greatest improvement in strength was noticed for subgroupB (Esthet-X) during empirical testing when the cavity design was altered (concave versus box). The conventional design (box shape) showed that Restofill had significantly better strength than Esthet-X (1- and 1/2-fold).

Cavities with large cavosurface angle (CSA), which are considered to lead to low stress concentrations near the free surface, were found to have better marginal integrity in a finite element study using axisymmetric models [19]. Similar cavity shapes were also investigated by Hembree Jr. [20]. A saucer-shaped cavity design was proposed and investigated for composite restorations [21–23]. In a clinical trial, 51 such preparations were completed and evaluated annually for 10 years and 70% of the restorations were acceptable for continued use [22]. A more recent study compared the survival of restorations placed in saucer-shaped cavities to that of restorations placed in tunnel preparations [23]. After a mean service life of 28.8 and 30.3 months, the proportion of the tunnel and saucer-shaped restorations survived was 46% and 76%, respectively. Moreover, saucer-shaped restorations showed lower caries development than the tunnel restorations after an observation period of 24 months. The alternative cavity shapes proposed in the above works were mostly based on experience and intuition of the investigators [18]. In the present study finite element modelling (FEM) and empirical testing showed that the conservative concave design, which had a larger CSA due to the 4° taper, demonstrated a two-and-half-fold increase in strength than the conventional box design which had a butt joint and hence lower CSA. Thus the results of the present agree with those of the above workers.

In another study using shape optimization technique, it was found that the optimized design showed significant (24%) improvement in the resistance against debonding under compressive load compared with the traditional design [24]. In the present study, the concave design (group II) showed 59% and 57% increase in compressive strength compared to conventional box design (group I) in FEM and empirical testing, respectively.

In a study on mechanics of fracture of enamel and dentin it was found that they behaved as brittle materials and that forces generated through mastication and bruxism had the potential to initiate fractures when the tooth had been altered by cavity preparation [25]. Therefore, the importance of designing the cavity preparation to reduce any potential stress concentrations was recognized. That radical removal of tooth structure results in reduction of fracture strength of teeth, and increased marginal gap formation has been well documented in many other studies [1, 26–29].

The results of the present study are in agreement with the above studies and the conservative design resulted in a stronger restoration. Even a material with a lower strength could perform equally well as a material with higher strength, if a conservative design is chosen for placing the restoration.

Hence the null hypothesis that preparation design does not affect the strength of direct composite restorations is rejected.

CONCLUSIONS

The present study tested the effect of cavity designs on only the compressive strength of direct composite restorations. The tensile and flexural strengths were not analyzed. Also only 2 composite materials were tested. Nevertheless the importance of conservative design has been highlighted by the findings of this study. Hence within the limitations of this invitro study, it may be concluded that a conservative concave cavity design with a 4° taper provides a significantly higher strength to direct composite restorations irrespective of the restorative material used. Such a cavity design also fulfills the objectives of minimally invasive dentistry. Hence the null hypothesis that design does not affect the strength of the direct composite restorations is rejected. However, further FEM analyses and clinical investigations of the other conservative cavity designs are needed to validate the alternate hypothesis that cavity design plays a significant role in improving the tensile and flexural strength of direct composite restorations.

ACKNOWLEDGMENTS

The authors wish to express their heartfelt sincere gratitude to Professor Krishnan Balasubramanian, head, Centre for Nondestructive Evaluation, Department of Mechanical Engineering, Machine Design

Section, Indian Institute of Technology Madras and Mr. Jothinathan, project consultant of the above laboratory for their kind help and support in completing the FE analysis.

REFERENCES

1. Fonseca RB, Fernandes-Neto AJ, Correr-Sobrinho L, Soares CJ. The influence of cavity preparation design on fracture strength and mode of fracture of laboratory-processed composite resin restorations.*Journal of Prosthetic Dentistry*. 2007; 98(4):277–284.
2. Porte A, Lutz F, Lund MR, Swartz ML, Cochran MA. Cavity designs for composite resins. *Operative Dentistry*. 1984; 9(2):50–56.
3. Wilson J. Effects of design features and restorative techniques on marginal leakage of MO composites: an in vitro study. *Operative Dentistry*. 1993; 18(4):155–159.
4. Bronner FJ. Engineering principles applied to cl II cavities. *Journal of Dental Research*. 1930; 10: 115–119.
5. Gabel AB. The amalgam proximoocclusal restoration in deciduous molars. *Journal of Dentistry for Children*. 1944; 11: 52–55.
6. Brown WB. A mechanical basis for the preparation of class II cavities for amalgam in deciduous molars.*Journal of the American Dietetic Association*. 1949;38: 417–423.
7. Weiland WF. An engineer's evaluation of Class II cavity preparations for silver amalgam restorations in primary teeth. *Journal of Dentistry for Children*. 1951;18(1):29–32.
8. Morin DL, Douglas WH, Cross M, Delong R. Biophysical stress analysis of restored teeth: experimental strain measurement. *Dental Materials*. 1988; 4(1):41–48.
9. Noonan MA. The use of photoelasticity in a study of cavity preparations. *Journal of Dentistry for Children*. 1949; 16(4):24–28.
10. Haskins RC, Haack DC, Ireland RL. A study of stress pattern variations in Class II cavity restorations as a result of different cavity design. *Journal of Dental Research*. 1954;33: 757–766.
11. Mahler DB, Peyton FA. Photoelasticity as a research technique for analyzing stresses in dental structures.*Journal of Dental Research*. 1955; 34(6):831–838.

12. Granath LE. Photoelastic model experiments on Class II cavity restorations of dental amalgam.*Odontologisk Revy*. 1965;16(supplement 9)

13. Craig RG, El-Ebrashi MK, LePeak PJ, Peyton FA. Experimental stress analysis of dental restorations. Part I. Two-dimensional photoelastic stress analysis of inlays. *The Journal of Prosthetic Dentistry*.1967; 17(3):277–291.

14. Granath LE, Hiltscher R. Strength of edge of class II cavity restorations in relation to buccolingual shape of cavity. *Odontologisk Revy*. 1970; 21(2):189–197.

15. Fisher DW, Caputo AA, Shillingburg HT, Duncanson MG. Photoelastic analysis of inlay and onlay preparations. *Journal of Prosthetic Dentistry*. 1975; 33(1):47–53.

16. Brekelmans WAM, Porte HW, Sloof TJJH. A new method to analyse the mechanical behaviour of skeletal parts. *Acta Orthopaedica Scandinavica*. 1972; 43: 301–317.

17. Vale WA. Cavity preparation. *Irish Dentist Reviews*. 1956; 2: 33–41.

18. Shi L, Fok ASL, Qualtrough A. A two-stage shape optimization process for cavity preparation. *Dental Materials*. 2008; 24(11):1444–1453.

19. Hubsch P, Middleton J, Knox J. The influence of cavity shape on the stresses in composite dental restorations: a finite element study. *Computer Methods in Biomechanics and Biomedical Engineering*.2002; 5(5):343–349.

20. Hembree JH., Jr. Microleakage of composite resin restorations with different cavosurface designs. *The Journal of Prosthetic Dentistry*. 1980; 44 (2):171–174.

21. Nordbo H, Leirskar J, von der Fehr FR. Saucer-shaped cavity preparation for composite resin restorations in class II carious lesions: 3 year results. *The Journal of Prosthetic Dentistry*. 1993; 69(2):155–159.

22. Nordbo H, Leirskar J, Von Der Fehr FR. Saucer-shaped cavity preparations for posterior approximal resin composite restorations: observations up to 10 years. *Quintessence International*. 1998; 29(1):5–11.

23. Horsted-Bindslev P, Heyde-Peterson B, Simonson P, Baelum V. Tunnel or saucer-shaped restorations: a survival analysis. *Clinical Oral Investigations*. 2005;9(4):233–238.

24. Li H, Yun X, Li J, et al. strengthening of a model composite restoration using shape optimization: a numerical and experimental study. *Dental Materials*. 2010; 26(2):126–134.

25. Rasmussen SE, Patchin RE, Scott DB, Heuer AH. Fracture properties of human enamel and dentin.*Journal of Dental Research*. 1976; 55(1):154–164.

26. Mondelli J, Steagall L, Ishikiriama A, de Lima Navarro MF, Soares FB. Fracture strength of human teeth with cavity preparations. *The Journal of Prosthetic Dentistry*. 1980; 43(4):419–422.

27. Malcolm PJ, Hood JAA. The effect of cast restorations in reducing cuspal flexibility of restored teeth.*Journal of Dental Research*. 1971; 56:p. D207.

28. Grimaldi JR, Hood JAA. Lateral deformation of tooth crown under axial and cuspal loading. *Journal of Dental Research*. 1973; 52:p. 584.

29. Soares CJ, Martins LRM, Fonseca RB, Correr-Sobrinho L, Fernandes Neto AJ. Influence of cavity preparation design on fracture resistance of posterior Leucite-reinforced ceramic restorations. *Journal of Prosthetic Dentistry*. 2006; 95(6):421–429.

Chapter 5

Balance Fatigue Design of Cast Steel Nodes in Tubular Steel Structures

Libin Wang,[1] Hui Jin,[2] Haiwei Dong,[2] and Jing Li[2]

[1]School of Civil Engineering, Nanjing Forestry University, Nanjing 210037, China

[2]Jiangsu Key Laboratory of Engineering Mechanics, Southeast University, Nanjing 210096, China

ABSTRACT

Cast steel nodes are being increasingly popular in steel structure joint application as their advanced mechanical performances and flexible forms. This kind of joints improves the structural antifatigue capability observably and is expected to be widely used in the structures with fatigue loadings. Cast steel node joint consists of two parts: casting itself and the welds between the node and the steel member. The fatigue resistances of these two parts are very different; the experiment results showed very clearly that the fatigue behavior was governed by the welds in all tested configurations. This paper focuses on the balance fatigue design of these two parts in a cast steel node joint using fracture mechanics and FEM. The defects in castings are simulated by cracks conservatively. The final crack size is decided by the minimum of 90% of the wall thickness and the value deduced by fracture toughness. The allowable initial crack size could be obtained through the integral of Paris equation when the crack propagation life is considered equal to the weld fatigue life; therefore, the two parts in a cast steel node joint will have a balance fatigue life.

INTRODUCTION

Due to their advanced mechanical performance, beautiful appearance, and smooth transitions, flexible forms that multiple steel tubes can intersect from different directions, cast steel nodes are especially applicable to joints of steel structures which are in the state of three-dimensional stress in civil engineering [1, 2]. The cast steel nodes were first introduced to the offshore structures in the 1980s [3]. From then on, many countries start using cast steel nodes in the design and construction of offshore structures. Numerous studies have been carried out on cast steel nodes in ocean engineering, and there are some reference standards now [4]. Nowadays, cast steel nodes also have been widely applied in civil engineering owning to its unique advantages. Compared with the traditional welded joints, cast steel nodes have some obvious advantages (see in Figures 1 and .2). (1) The girth butt welds are used which are perpendicular to the axis of the tube, so that the welding secondary stress can be reduced greatly. (2) The welded seams are far from the node core area to reduce the weld stress. (3) By chamfering at tube intersections, the stress concentration is reduced. (4) The casting production process has great applicability and can meet various requirements of appearance and mechanics. (5) Cast steel node is good at fatigue resistance and corrosion resistance.

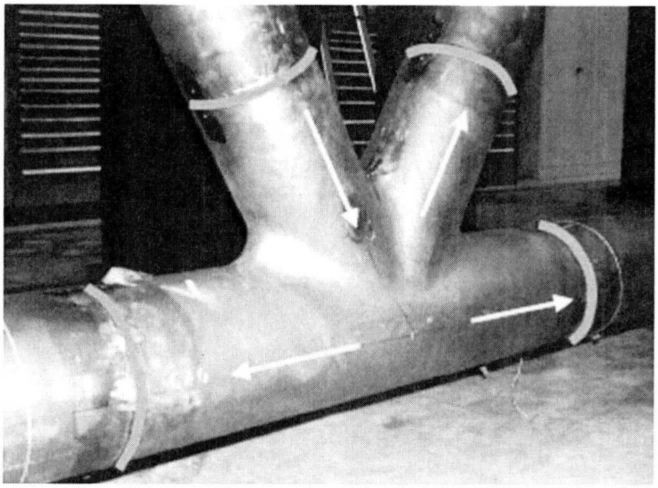

Figure 1: Cast steel node (Pictures are from http://icom.epfl.ch/).

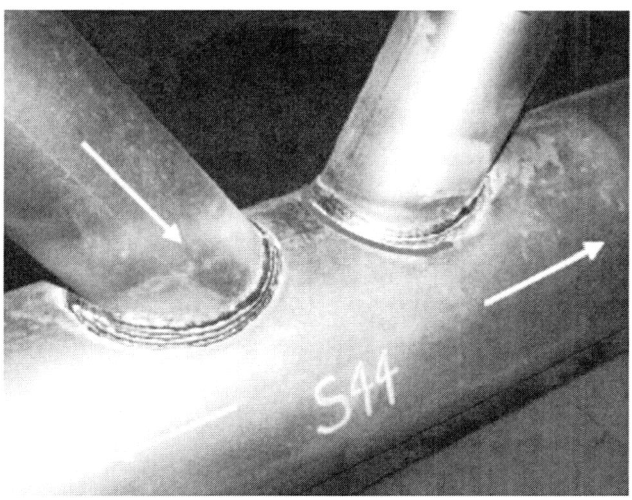

Figure 2: Welded joint (Pictures are from http://icom.epfl.ch/).

The laboratory of ICOM in Switzerland did fatigue tests on large-scale steel tube trusses containing cast steel nodes. They studied the fatigue properties of the cast steel nodes and the whole trusses, the influence of casting defects, and compared the fatigue properties of cast steel node with that of traditional welded node [5–9]. The researches in Germany studied and tested the fatigue strength of the girth butt welds in cast steel node [10]. Jin et al. studied the fatigue strength of castings and the girth butt welds in cast steel nodes in China [11–14].

The fatigue capacity of cast steel nodes is decided by two major parts: the casting itself and the girth butt welds between the casting and the steel members. The fatigue resistances of these two parts are very different; the experiment results showed very clearly that the fatigue behavior was governed by the welds in all tested configurations [7, 9]. So, this kind of joint contains two greatly unbalanced parts from an overall fatigue perspective. The balance fatigue design can realize an optimization of performance and economy.

The casting defects are unavoidable in the process of casting. If defects are permitted, there must remarkably influence the fatigue property of casting part, as cracks are initiated from defects. If defects are not permitted, there will be lots of fabrication difficulties, more restrictions on the node form, and higher probabilities of unacceptable quality production. Moreover, the additional fatigue strength of the casting material exceeding the welds is wasted. The key point of balance design is to calculate the allowable initial defect size and make the casting fatigue strength containing defects coordinated with the weld fatigue strength. Thus, it could reach a best economic efficiency by optimum using of the materials. The damage tolerance method is used in this paper to quantitatively calculate the allowable initial defects size in castings for a balance fatigue design.

METHOD OF DAMAGE TOLERANCE DESIGN

Method of damage tolerance design is based on the fracture mechanics theory, nondestructive test, and fracture toughness test, which is firstly applied in the fatigue balance design of cast steel node by Nussbaumer and Haldimann-Sturm [7, 9].

Stress Intensity Factor

In fracture mechanics, the stress intensity factor (SIF) expression is

$$K = f\sigma\sqrt{\pi a}, \tag{1}$$

In which f is correction factor decided by the crack shape, crack location, and loading mode. f may be a constant as it is proposed in [7, 9], or in function of the crack size along different possible crack propagation directions as it is fitted in this paper by FEM simulation results.

Fatigue Crack Growth Rate

Fatigue crack growth rate can be expressed by Paris-Erdogan equation:

$$\frac{da}{dN} = C(\Delta K)^m, \tag{2}$$

In which K is SIF range, $K = K_{max} - K_{min}$; C, m are material constants.

Estimation of Residual Life

Initial Crack Size a_0

The initial crack size has great influence on the fatigue life and should be prudently decided [15]. The casting defects are unavoidable in the process of casting like gas holes, slack inclusions, or shrinkages. These defects can be modeled as two-dimensional cracks, and this is a conservative assumption [7, 9]. There are two kinds of normalized cracks in castings: surface cracks and inner cracks. The key point of balance design is to calculate the allowable initial defect size a_0, making the casting fatigue strength containing defects coordinated with the weld fatigue strength.

The real initial crack size a_0' can either be evaluated by casting process numeral simulation or be detected by nondestructive testing.

If a_0' is smaller than the assessed allowable initial defect size a_0, the fatigue life is still governed by the weld.

Critical Crack Size a_c

Critical crack size is the maximum size allowed under certain stress condition, in the case of brittle failure. It is often expressed by a_c, and determined by the fracture toughness K_{Ic}:

$$a_c = \frac{1}{\pi}\left(\frac{K_{Ic}}{f\sigma}\right)^2 (mm). \tag{3}$$

The final crack size for castings is decided by the minimum of 90% of the wall thickness as a through-thickness crack and the value calculated by fracture toughness (see (3)).

Fatigue Crack Propagation Life

The fatigue crack propagation life is deduced through the integral of the Paris-Erdogan equation as the constant-amplitude stress is considered:

$$N_P = \int_{N_0}^{N_f} dN = \int_{a_0}^{a_c} \frac{da}{C(f\Delta\sigma\sqrt{\pi a})^m}. \tag{4}$$

If f is the function of a, the $(a_c - a_0)$ can be divided into several intervals, f and N_{pi} can be calculated with the mean value of a in each interval. Then, by summing the N_{pi} of each interval, the fatigue crack propagation life can be evaluated.

Balance Fatigue Design of Cast Steel Nodes

Because of the defects existing, the fatigue mechanical performance of castings is lowered. The defect size and location can be detected by nondestructive testing method after casting fabrication. So, the key point is to investigate and evaluate the influence degree and how much the residual strength is.

This section focuses on the design process of cast steel nodes with initial defects, proposing the method of optimization and balance design. The main procedures are as follows.

1. *The Node Shape Design.* The node shape can be designed preliminarily according to the mechanics requirements from the whole structure force-bearing demand. and then may be modified according to the pouring and casting process technical feasibility.
2. *Optimization Design of the Girth Butt Welds.* The fatigue capability of the joint is governed by the welds, so the girth butt welds should be designed optimally. The hot spot stresses should be analyzed on different weld details, and the optimum weld design detail can be decided by the hot stress life curve; therefore, the weld life N can be calculated as the whole joint life which will be used to evaluate the allowable initial crack size a_0.
3. *Casting Quality Level Determination.* Higher casting quality cost more; there will be lots of fabrication difficulties, more restrictions on the node form, and higher probabilities of unacceptable quality production. Moreover, the additional fatigue strength of the casting material exceeding the welds is wasted, so reasonable casting quality requirements can balance the strengths of the two parts in one joint. According to the joint fatigue life N, the stress amplitude in the casting part, and the material S-N curves at different cast quality level, the reasonable casting quality levels can be determined.
4. *Real Initial Crack Size Determination.* During the design phase, the real initial crack size a_0' can be evaluated by pouring and casting process numeral simulation through some professional software.
5. *Justification of a_0'.* The casting defects can be modeled as cracks and an equivalent initial crack size $a_0$represents the initial defect size. The determination of an allowable a_0 value can be deduced by fracture mechanics, when the critical crack size a_c is determined and the life is equal to the joint life N. Then, the real Initial crack size a_0' is compared with a_0; if it is greater than the allowable size, the design should be modified.

The procedures previously mentioned provide a basis for balance design and production processes of cast steel node, and the flowchart is shown in Figure 3.

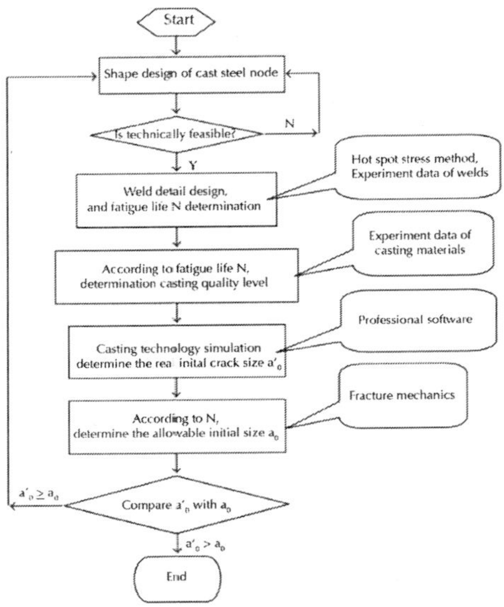

Figure 3: Procedures of optimization design of cast steel nodes.

Determination of Allowable Initial Size a_0

Nussbaumer and Haldimann-Sturm proposed the correction factor *f* is constant [7, 9]; in this study it is found a variable not only in function of *a*, but also in function of propagation direction:

$$K_i = f_i\left(\frac{a}{t}\right)\sigma\sqrt{\pi a}, \tag{5}$$

in which K_i is SIF along propagation direction *i*. The crack is modeled by FEM, ANSYS software, and f_i is fitted to get the K_i expression.

The SIF range is simulated by FEM

$$\Delta KI(a) = f.\Delta\sigma.\sqrt{\pi.a}, \qquad (6)$$

In which $= _{max} - _{min}$.

Determination of allowable initial defect size a_0 is in Figure 4.

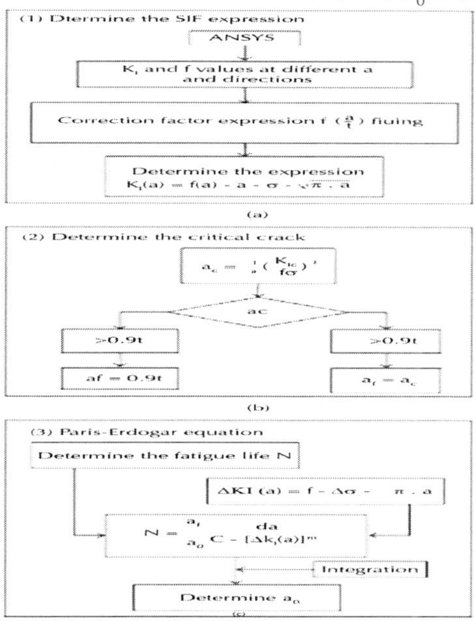

Figure 4: Calculation of initial crack size.

Examples

The design of cast steel node in tour tower steel structure on the platform of Hangzhou Bay Bridge is demonstrated to show the balance design procedures.

Five typical defects were found in the castings, in which ?, ?, and ? are modeled as inner cracks and ?, ? are modeled as surface cracks (Figure 5). Due to the longitudinal symmetry, half model is analyzed;

the cracks are modeled as a half of circular crack and a quarter of circular crack. The defect locations are shown in Figure 5. The designed real initial crack size a_0' are listed in Table 1.

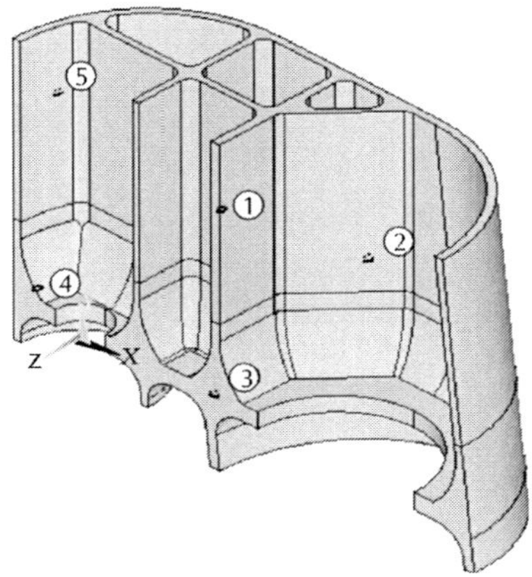

Figure 5: Node model and defect locations.

Table 1: Real initial defect size a_0' (mm).

Defect					
a_0'	15.4	7.6	29.5	21.6	10.8

Modeling the Cracks

The surface crack body of defect ⍰ is modeled by a quarter of one thin circle cylinder due to the symmetry (Figure 6), consisting of two parts: upper part and lower part to construct the crack surface and crack tip. The crack body meshes are refined (Figure 7). The inner crack body of defect ⍰ is modeled by a half of one thin circle cylinder (Figure 8), also consisting of two parts: upper part and lower part to construct the crack surface and crack tip. The crack body meshes are refined (Figure 9).

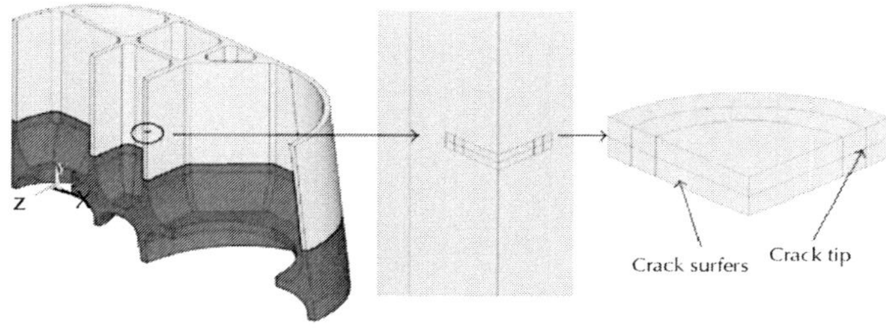

Figure 6: Surface crack body.

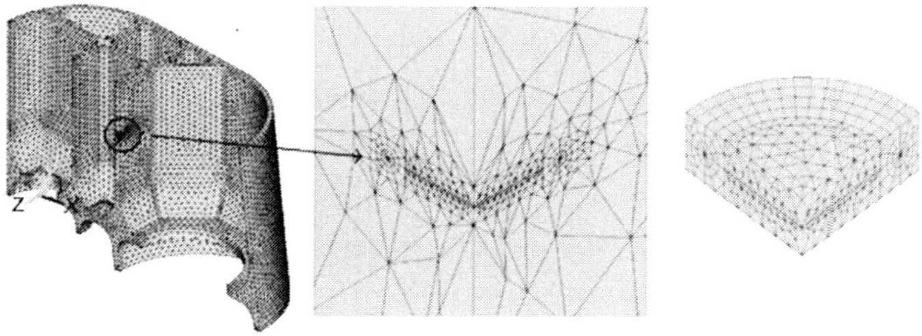

Figure 7: Refined surface crack body meshes.

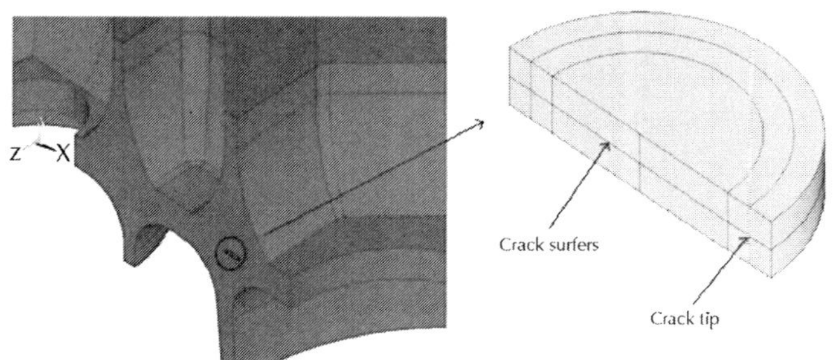

Figure 8: Inner crack body.

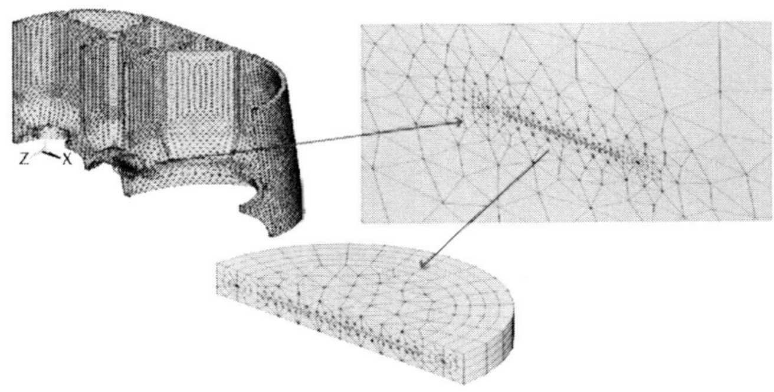

Figure 9: Refined inner crack body meshes.

Cyclic Loadings

The constant-amplitude cyclic loadings inducing fatigue on the node are shown in Figure 10, Table 2, consisting of axis force *P*, sheering force *V*, and bending moment *M* [16].

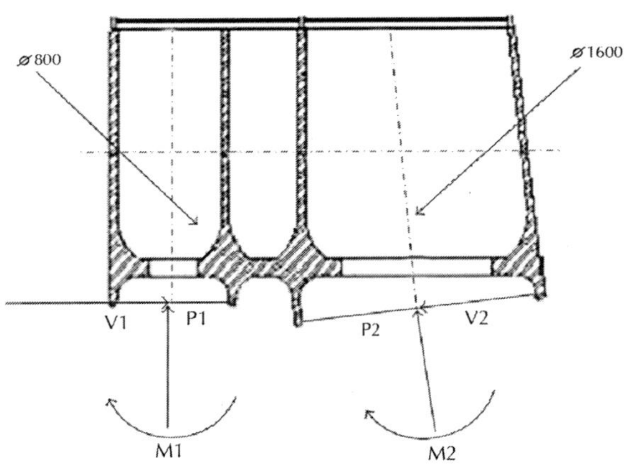

Figure 10: Loads on node.

Table 2: Constant amplitude fatigue loads

	Load case		P (KN)	V (KN)	M (kN · m)
On Tube of Ø 800	Case 1	Max	756.391	24.132	15.5521
	Case 2	Min	−736.617	−23.456	−15.1126
On Tube of Ø 1600	Case 1	Max	2000.222	49.266	209.1114
	Case 2	Min	−1956.73	−48.601	−214.845

Calculation of Allowable Crack Size

Defects ⍰ and ⍰ are chosen to demonstrate the calculation procedures of allowable initial crack size of surface and inner defect.

Calculation of Allowable Initial Surface Crack Size of Defect ⍰

The principal stress contour of the crack body is shown in Figure 11.

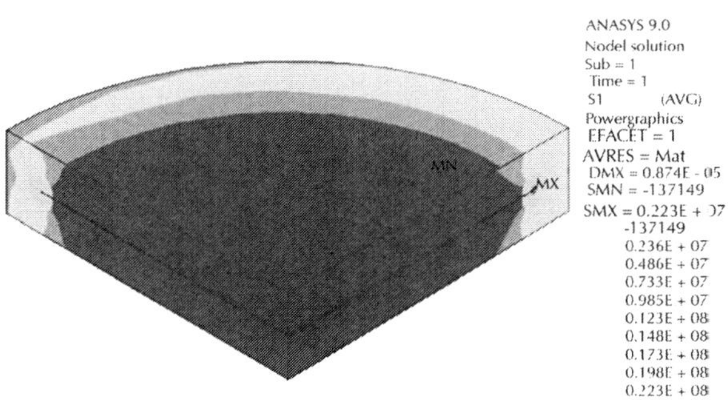

Figure 11: Principal stress contour of crack body ⍰.

Defect ⍰ locates at the edge of inner stiffener plate of the node, where the node wall thickness is t = 45 mm. Five cases of different crack depth are analyzed by ANSYS, that is, a = 15 mm, 20 mm, 25 mm, 30 mm, and 35 mm in each case the crack body is meshed along sixteen prorogation directions as symbol , that is, = 0°, 6°, …. , 90°. See Figure 12.

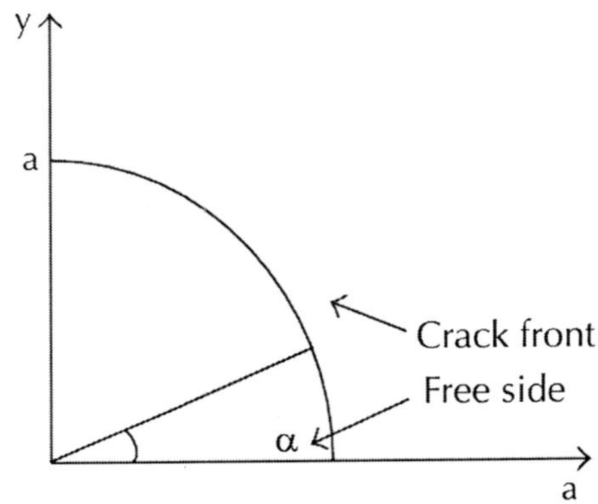

Figure 12: Prorogation directions.

In ANSYS, the crack propagating path should be defined, a crack path is defined by nodes 1, 2, 3, 4, 5, for instance (Figure 13). The crack SIF is calculated as the average of SIF values at the nodes on the crack path.

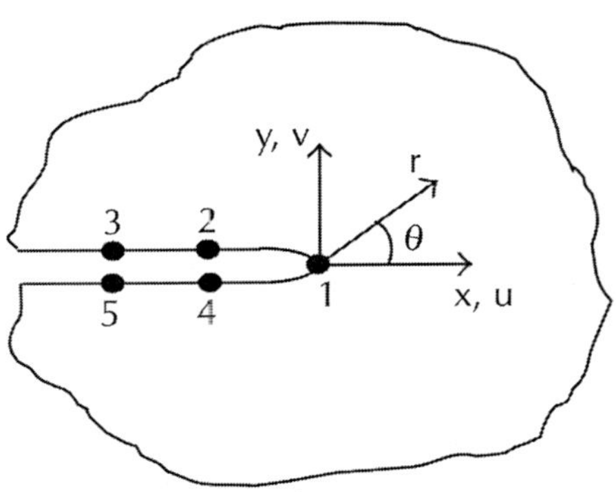

Figure 13: Crack path defined in ANSYS.

The expression of K_I is [17]

$$K_I = \frac{E'}{4}\sqrt{\frac{\pi}{2L}}\left[4(v_2 + v_4) - v_3 - v_5\right], \tag{7}$$

In which $E' = E/(1 - \mu^2)$ (plane strain), E is elastic modulus, μ is passion ratio, is node displacement on the crack path, and L is the length of the crack tip element.

SIFs of five cases along sixteen direction path are shown in Figure 14, and a curve is fitted for each case to show the trend. The SIF values are obviously not constant; their changing ratio ranges are 63% ~ 67% along the direction from = 0° to = 90° and 26% ~ 410% along the crack depth from a = 15 mm to a = 35 mm.

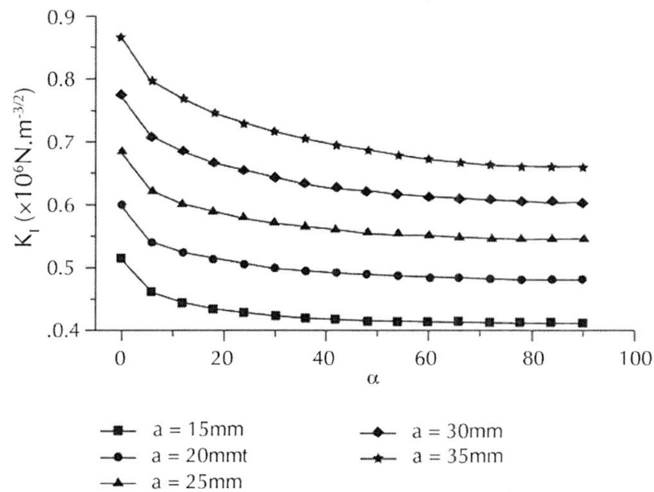

Figure 14: SIF fitting of defect .

The path of maximum SIF value is considered where = 0° (see in Figure 14). Along the path of = 0°, the values of f at five crack depths are calculated through (5), and a quartic polynomial is fitted (Figure 15).

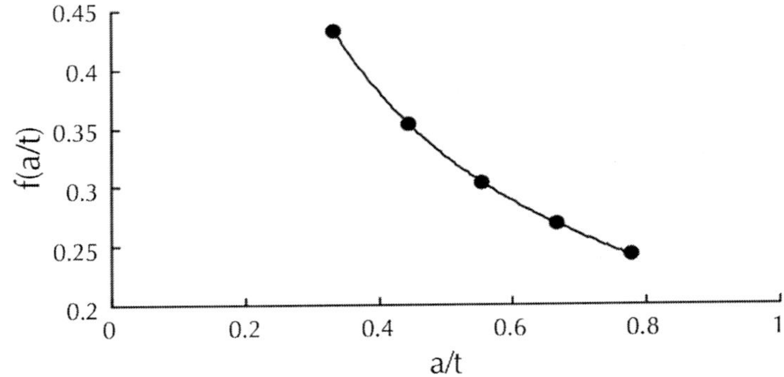

Figure 15: Fitting curve of *f* at the path of = 0° .

The expression of quartic polynomial is

$$f\left(\frac{a}{t}\right)=1.1486\left(\frac{a}{t}\right)^4-3.7431\left(\frac{a}{t}\right)^3+4.7291\left(\frac{a}{t}\right)2-2.9461\left(\frac{a}{t}\right)+1.0127 \tag{8}$$

By substitution of the data of wall thickness t = 45 mm, the expression of *K* in function of *a* can be obtained through *f* fitting,

$$K_I=\sigma\sqrt{\pi}\begin{pmatrix}280103.6427a^{4.5}-41076.5432a^{3.5}+2335.3580a^{2.5}\\-65.4689a^{1.5}+1.0127a^{0.5})\end{pmatrix} \tag{9}$$

The casting material used in the steel tower structure at Hangzhou Bay is GS-20Mn5V, and there is no fatigue and fracture data available for reference. As the element compositions and mechanical properties of GS-Mn5V are similar to ZG20SiMn (Chinese Steel Grade) [18], the material parameters of ZG20SiMn in [19] are used approximately: $C = 2.25510^{-13}$, m = 3.9917. The critical J integral of ZG20SiMn is J = 102.6 (N/mm), so the fracture toughness can be deduced in the case of plane strain

$$K_{IC}=\sqrt{j.\frac{E}{1-\mu^2}}=152.4\left(MP\sqrt{m}\right). \tag{10}$$

The allowable initial crack size of defect ⍰ is calculated following the flowchart (Figure 4) with the FEM analysis results; the result is a_C = 27.6 mm, and the designed real crack size is a_0' = 15.4 mm, so the casting strength is satisfied.

Calculation of Allowable Initial Inner Crack Size of Defect ⍰

The principal stress contour of the crack body is shown in Figure 16.

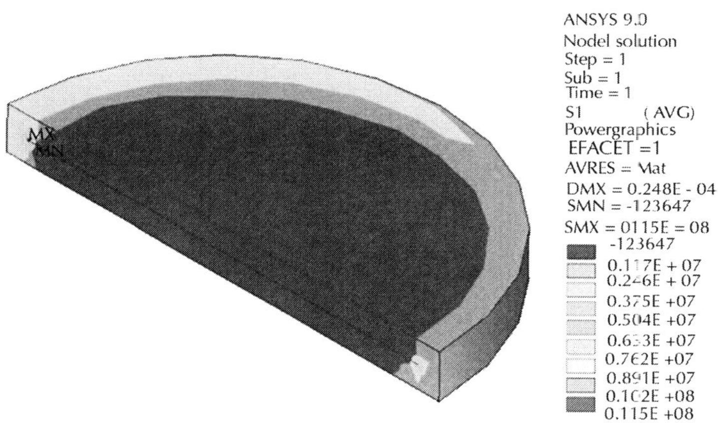

Figure 16: Principal stress contour of crack body ⍰.

Defect ⍰ locates at the inner of the node, where the horizontal node thickness along the crack propagation direction is t = 800 mm. Five cases of different crack depth are analyzed by ANSYS, that is, a = 25 mm, 30 mm, 35 mm, 40 mm, and 45 mm. In each case, the crack body is meshed along thirteen prorogation directions as symbol , that is, = 0°, 15°, ..., 180°. See Figure 17.

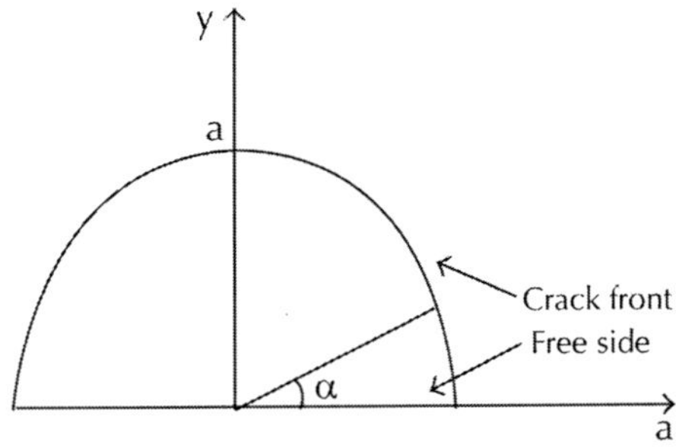

Figure 17: Prorogation directions.

SIF values of five cases along thirteen direction paths are shown in Figure 18, and a curve is fitted for each case to show the trend. The SIF value changing ratio ranges are 21% ~ 44% along the direction from $\alpha = 0°$ to $\alpha = 180°$, and 23% ~ 47% along the crack depth from $a = 25$ mm to $a = 45$ mm.

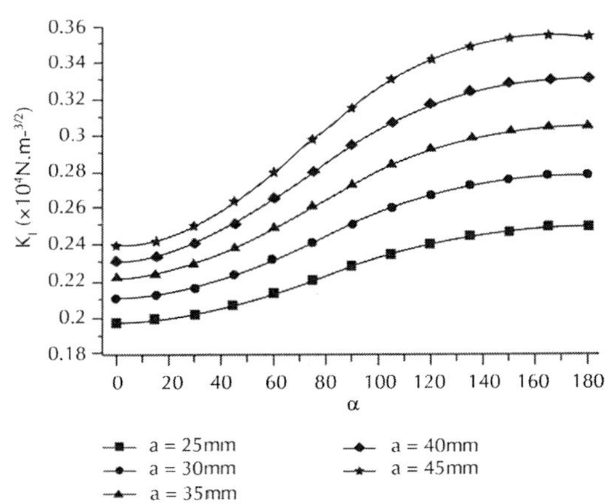

Figure 18: SIF fitting of defect ?.

The path of maximum SIF value is considered where $\alpha = 180°$ (see in Figure 18). The fitting quartic polynomials are shown in Figure 19.

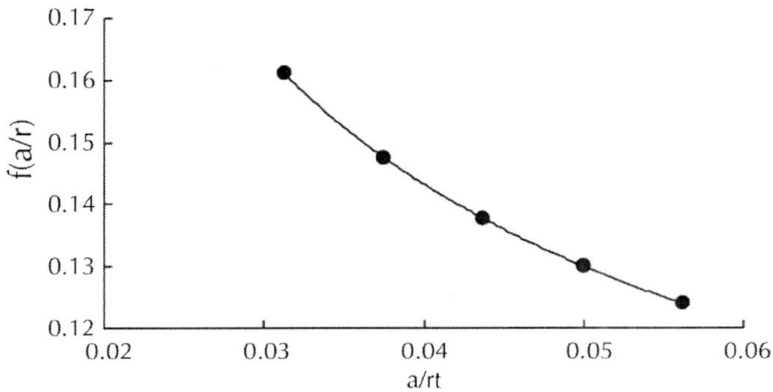

Figure 19: Fitting curve of *f*at the path of $\alpha = 180°$.

One has

$$f\left(\frac{a}{t}\right) = 30037\left(\frac{a}{t}\right)^4 - 5905.1\left(\frac{a}{t}\right)^3 + 457.07\left(\frac{a}{t}\right)^2 - 17.529\left(\frac{a}{t}\right) + 0.4142 \qquad (11)$$

By substitution of the data of thickness $t = 800$ mm, the expression of K in function of a is

$$K_I = \sigma\sqrt{\pi}\left(73332.5195a^{4.5} - 11533.4a^{3.5} + 714.2a^{2.5} - 21.9a^{1.5} + 0.4142a^{0.5}\right) \qquad (12)$$

the allowable initial crack size of defect ⍰ is calculated following the flowchart (Figure 4) with the FEM analysis results, the result is a_0 = 221.3 mm, and the designed real crack size is a_0' = 29.5 mm, so the casting strength is satisfied.

RESULTS OF FIVE DEFECTS

The results of five defects are listed in Table 3, the ratio ranges of allowable initial defect size to wall thickness are from 28% to 67%, the max-ratio happens at the thinnest wall thickness location, and the min-ratio happens at the thickest inner node thickness location. The minimum allowable initial defect locates at defect ⍰, and the maximum allowable defect locates at defect.

Table 3: Results of five defects.

Defect	Crack path	Wall thickness t (mm)	α_0 (mm)	α_0' (mm)	α_0/t (%)
	$\alpha = 0°$	45	27.6	15.4	61
	$\alpha = 90°$	35	23.3	7.6	67
	$\alpha = 180°$	800	221.3	29.5	28
	$\alpha = 0°$	86	37	21.6	43
	$\alpha = 180°$	45	17.9	10.8	40

It is shown that the present design with five defects satisfies the strength demand, and the results and conclusions can be referenced for further design improvement.

CONCLUSIONS

This paper focuses on the balance fatigue design of the two parts in a cast steel node joint using fracture mechanics and FEM; the key point in the procedures is how to calculate the allowable initial equivalent crack size; the following main conclusions can be drawn.

1. The allowable existence of reasonable initial defects make the castings fatigue strength coordinated with the weld fatigue strength to realize a balance design avoiding the material strength waste.
2. Casting defects can be equivalently treated as cracks and simulated by FEM software fracture module.
3. The engineering example shows the ratio of allowable initial defect size to node thickness ranges from 28% to 67%, in which the high ratio may not be allowed without this balance design concept.
4. The crack SIF values shown in the fracture mechanic simulation results are obviously not constant; it depends on the crack depth and crack propagation direction and their changing ratio can be up to 410%; the SIF accuracy will seriously affect the fracture

results, so whether the SIF can be simplified as a constant correction factor or not should be determined carefully by a given requisite degree of accuracy.

5. The node has distinct different fatigue capabilities under some influences, for example, locations, directions, and loading conditions. For the fatigue sensitive ones, it is important to estimate the defect propagating direction and the range of SIF K correctly.

CONFLICT OF INTERESTS

The authors declare that there is no conflict of interests regarding the publication of this paper.

ACKNOWLEDGMENTS

The work reported is supported by National Natural Science Foundation of China (51108075), The Natural Science Foundation of Jiangsu Province (BK2011613), Science and Technology Project of the Ministry of Housing and Urban-Rural Development (2010-K2-7), and A Project Funded by the Priority Academic Program Development of Jiangsu Higher Education Institutions.

REFERENCES

1. de Oliveira JC, Willibald S, Packer JA, Christopoulos C, Verhey T. Cast steel nodes in tubular construction—Canadian experience. Proceedings of the 11th International Symposium and IIW International Conference on Tubular Structures; September 2006; Quebec City, Canada. pp. 523–529.
2. Schober H. Steel castings in architecture and engineering. *Modern Steel Construction*. 2003; 43(4):65–72.
3. Ma A, Sharp JV. Fatigue design of cast steel nodes in offshore structures based on research data. Proceedings of the Institution of Civil Engineers, Water Maritime and Energy; June 1997; pp. 112–126.

4. DNV-RP-203 fatigue design of offshore structures, April 2010.
5. Nussbaumer A. Defect tolerances of steel castings in bridge structures. *Report Scientifique, ICOM Project*.2001–2005;(IC 438)
6. Schumacher A, Sturm S, Walbridge S, Nussbaumer A, Hirt MA. *Rapport ICOM*. 489E. Lausanne, Switzerland: ICOM Laboratoire de la construction métallique; 2003. Fatigue design of bridges with welded circular hollow sections.
7. Nussbaumer A, Haldimann-Sturm SC, Schumacher A. Fatigue of bridge joints using welded tubes or cast steel node solutions. Proceedings of the 11th International Symposium and IIW International Conference on Tubular Structures; September 2006; Quebec City, Canada. pp. 56–63.
8. Haldimann-Sturm S, Nussbaumer A. Tubular bridges and the fatigue design of their joints. Proceedings of the Eurosteel Conference on Steel and Composite Structures; June 2005; pp. 45–52.
9. Haldimann-Sturm SC, Nussbaumer A. Fatigue design of cast steel nodes in tubular bridge structures.*International Journal of Fatigue*. 2008; 30(3):528–537.
10. Veselcic M, Herion S, Puthli R. Selection of butt-welded connections for joints between tubulars and cast steel nodes under fatigue loading. Proceedings of the 11th International Symposium and IIW International Conference on Tubular Structures; September 2006; Quebec City, Canada. pp. 528–535.
11. Jin H, Li J, Li A. Checking calculation of fatigue strength for cast steel joints of offshore tour tower under wave loads. *Journal of Southwest Jiaotong University*. 2010; 45(5):692–697.
12. Jin H, Li J, Zhang Q, Wang H. Application of hot spot stress method to fatigue life evaluation of girth butt welds of cast steel node. *Journal of Tongji University*. 2009; 37(1):20–25.
13. Jin H, Zhang Q, Wang H. Check on fatigue strength of members for cast steel node. *Steel Construction*.2008; 23(112):17–21.
14. Ge J, Jin H. Fatigue life evaluation and detail design for girth butt welds of cast steel node. *Steel Construction*. 2008; 23(112):26–31.

15. Yao W. *Analysis of Structural Fatigue Life*. Peking, China: National Defence Industry Press; 2003. (Chinese)

16. Li J. *Fatigue properties and optimization design of cast steel nodes [M.S. thesis]* Nanjing, China: Mechanic Engineering Major in Southeast University; 2010. (Chinese)

17. Xie D, Qian Q, Li C. *Numerical Calculation Method of Fracture Mechanics and Its Application in Engineering*. Peking, china: Science press; 2009. (Chinese)

18. Yang Z, Shen L. Research and practice of welding technology of GS-20Mn5V cast steel node. *Journal of Modern Welding*. 2004;57(9):45–46.

19. Manual of Mechanical Engineering Material Properties Editorial board. *Manual of Mechanical Engineering Material Properties*. Peking, China: Machinery Industry Press; 1994. (Chinese)

Chapter 6

Strut Deformation in CFRP-Strengthened Reinforced Concrete Deep Beams

Mohammad Panjehpour[1], Hwa Kian Chai[1], and Yen Lei Voo[2]

[1]Department of Civil Engineering, Faculty of Engineering, University of Malaya, 50603 Kuala Lumpur, Malaysia

[2]DURA Technology Sdn Bhd, 31200 Chemor, Perak, Malaysia

ABSTRACT

Strut-and-tie model (STM) method evolved as one of the most useful designs for shear critical structures and discontinuity regions (D-regions). It provides widespread applications in the design of deep beams as recommended by many codes. The estimation of bottle-shaped strut dimensions, as a main constituent of STM, is essential in design calculations. The application of carbon fibre reinforced polymer (CFRP) as lightweight material with high tensile strength for strengthening D-regions is currently on the increase. However, the CFRP-strengthening of deep beam complicates the dimensions estimation of bottle-shaped strut. Therefore, this research aimed to investigate the effect of CFRP-strengthening on the deformation of RC strut in the design of deep beams. Two groups of specimens comprising six unstrengthened and six CFRP-strengthened RC deep beams with the shear span to the effective depth ratios (*a/d*) of 0.75, 1.00, 1.25, 1.50, 1.75, and 2.00 were constructed in this research. These beams were tested under four-point bending configuration. The deformation of struts was experimentally evaluated using the values of strain along and perpendicular to the strut centreline. The evaluation was made by the comparisons between unstrengthened and CFRP-strengthened struts regarding the widening and shortening. The key variables were *a/d* ratio and applied load level.

INTRODUCTION

Strut-and-tie model (STM) method evolved as one of the most useful designs for shear critical structures and discontinuity regions (D-regions). It reduces complex states of stress within the D-region in reinforced concrete (RC) member into a truss comprised of simple and uniaxial stress paths. The concrete stress field in strut, as a main constituent of STM, is usually wider at midlength of the strut than at the ends. Thus, strut generally varies in cross-section along its length. The strut that changes in its width along the length is idealised as a bottle-shaped geometry [1–7].

The cracking behaviour of bottle-shaped strut resembles that of concrete cylinder in a split tensile test. The internal lateral spread of applied compression force results in transverse tension strain in the strut which causes split cracking [2]. The bottle-shaped unreinforced strut longitudinally cracks while loading and it may cause ultimate failure of strut. Carbon fibre reinforced polymer (CFRP) has widespread applications for repair and strengthening of cracked concrete structural elements [8, 9]. The load capacity of strut can be increased by transverse reinforcement steel bar and external reinforcement of carbon fibre reinforced polymer (CFRP) sheet which restrain cracking. Dimensioning of strut, tie, and nodes is requisite to design the STM after computation of reactions and internal forces [10, 11]. Nevertheless, CFRP-strengthening of RC struts complicates the aforesaid dimensioning as CFRP-strengthening restrains the spread of stress fields.

STM method has been used for D-region analysis in structural elements such as deep beams according to many codes and standards [1–6]. Deep beams have been commonly used for high-rise buildings, offshore structures, and foundations [12], which are largely taken into account as a transfer girder in a single-span or continuous beam [7]. Experimental tests on reinforced concrete (RC) deep beams revealed that addition of web reinforcement beyond the recommended amount by design codes and standards was incapable of increasing the shear strength of deep beam [13]. Thus, CFRP sheet may be regarded as a proper material for external strengthening of D-regions in RC deep beams.

Prior research showed that the CFRP-strengthening increased the shear strength of RC deep beams in the range of 24–43% depending on the installation style and fibre direction of CFRP sheet [13]. This range was 35–73% for RC deep beams with the opening [14] and 10–66% for RC T-section deep beams [15]. The range of increase depended on the size of opening and CFRP-strengthening length, respectively. The strut effectiveness factor in STM recommended by ACI 318-05 was later modified for deep beam which was longitudinally reinforced by FRP bars with shear span to effective depth (a/d) ranging in $1.5 < a/d < 2.5$ [16] and $1.4 < a/d < 2.1$ [17]. To date, no research has been conducted on the effect of CFRP sheet bonding on RC strut behaviour, particularly the strut dimensions in STM of CFRP-strengthened D-region.

There are uncertainties regarding strut dimensioning and deformation while it is subject to load. Width of a strut has been estimated in the previous studies with the assumption that the strut has been stressed to the maximum [7, 18–21]. The dimensions estimation of bottle-shaped strut is essential in design calculations as the mechanical behaviour of strut is mainly influenced by its dimensions [2, 7, 19]. Among the latest publications on the STM and CFRP-strengthened RC deep beams [22–24] no research focused on the strut deformation subjected to load. Hence, determination of strut deformation under loading is vital for a reliable design. Furthermore, the strut deformation provides indication to the strut behaviour and subsequently to the overall behaviour of RC structural element.

The objective of this research is to investigate the effect of CFRP-strengthening on the deformation of RC strut in deep beam as it affects the failure of strut. The bonding between CFRP sheet and strut disrupts the load trajectories in strut and complicates the estimation of strut dimensions while loading. Hence, this investigation sets the scene for STM of CFRP-strengthened D-regions regarding strut dimensioning.

METHODOLOGY

Tests were performed on two groups of six RC deep beam specimens with and without CFRP-strengthening. Each group of beams had the shear span to the effective depth ratios of 0.75, 1.00, 1.25, 1.50, 1.75, and 2.00. Both groups of beam specimens were identical in every aspect, excluding the condition of strengthening by CFRP sheets. The variables of the investigation were *a*/*d* ratio and the applied load level as these two variables affect the deformation of struts and mechanical behaviour of RC deep beams.

A typical bottle-shaped strut is illustrated in Figure 1. Bulging stress trajectories generates considerable transverse stress in the midlength of strut with compression in the neck and tension further away leading to longitudinal cracking. The dashed outline in Figure 1 indicates approximately the boundaries of bottle-shaped strut. The bottle-shaped strut consists of other struts and ties as shown in Figure 1 using dashed line and solid line, respectively [2].

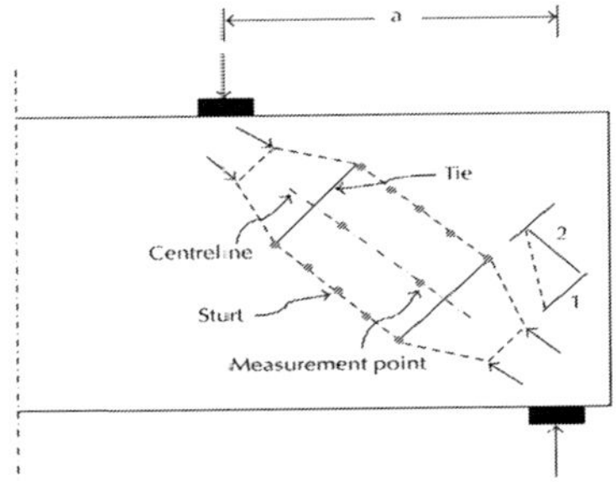

Figure 1 Strut-and-tie model of a bottle-shaped strut including measurement points.

The CFRP sheet was utilised in the current experiment as an external reinforcement to restrain cracking of RC struts in deep beams. The outline of bottle-shaped strut shown in Figure 1 is mainly based on the load path from support plate to load plate. The deformation of strut in unstrengthened and CFRP-strengthened D-region was evaluated by measuring the average values of strains along and perpendicular to the strut centreline. A demountable mechanical strain gauge (DEMEC) was utilised for strain measurement. The shear failure of deep beams was dominant in this experiment. The shortening and widening of the strut were measured in each step of loading with an increment of 50 kN till failure of strut.

Deep Beams Details

All the beam specimens were 1840 mm in length, 140 mm in width, and 350 mm in height. Nine deformed steel bars of 16 mm diameter were longitudinally placed in three layers in the bottom of beams as flexural reinforcement. Orthogonal grids of reinforcement with the spacing of 100 mm were provided at two sides of beams using deformed steel bars of 6 mm diameter. The longitudinal steel bars were welded to the two end steel plates to provide adequate anchorage. The end steel plates with 120 mm of height and 10 mm of thickness covered the width of beams fully at both ends of beams. The additional reinforcements (steel cage) using deformed steel bars of 6 mm diameter were placed under the load plates and atop the support plates to prevent local bearing stress. The beams details are indicated in Figure 2.

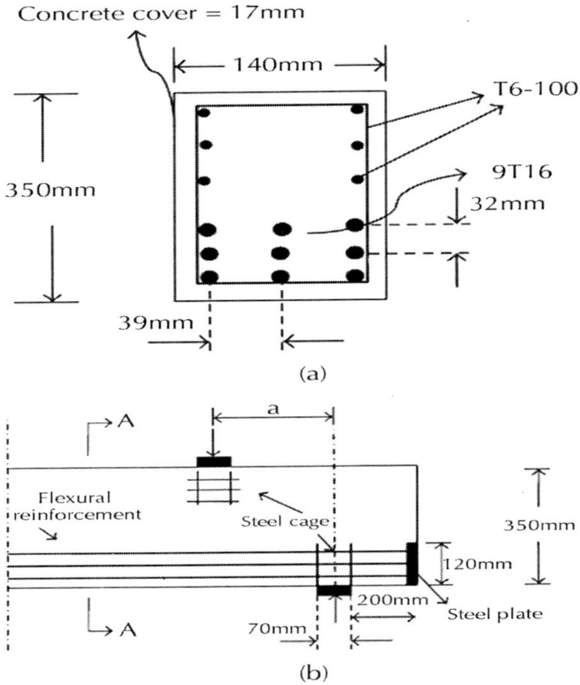

Figure 2 Beam details for half length: (a) section A-A, (b) side view.

Material and Specimens Preparation

Ordinary Portland concrete with 28-day cylindrical compressive strength of 37.02 MPa and splitting tensile strength of 3.31 MPa was used to fabricate all the beams. The maximum aggregate size and water to cement ratio were 10 mm and 0.48 in the concrete mix design, respectively. The density of concrete was 2420 kg/m^3. One layer of unidirectional CFRP sheet with the thickness of 0.111 mm/ply and two-part epoxy impregnation resin were utilised for strengthening. The rectangular surface areas between load and support plates (D-region) were fully covered by CFRP sheet at two sides of beams using wet lay-up system as shown in Figure 3. Table 1 presents the typical properties of the CFRP sheet and epoxy resin provided by the manufacturer. The CFRP sheet and epoxy resin were supplied by Sika Company with product data sheet of sikawrap-230C and sikadur-330, respectively. The required curing time for CFRP sheet after bonding was two days at ambient temperature as recommended by the manufacturer. The support plates and the load plates, which were 70 mm in width and 10 mm in thickness, fully covered the bottom and top of the beams. All the beams were cured for two weeks after casting with a wet mat.

Figure 3: Test setup.

Table 1: Typical properties of CFRP sheets and epoxy

Materials	Tensile strength (MPa)	Tensile modulus of elasticity (GPa)	Bond strength (MPa)	Fabric thickness (mm/ply)	Laminate thickness impregnated with resin (mm/ply)
CFRP sheet	3900	230	—	0.111	0.9
Epoxy resin	30	4.5	>4	—	—

Instrumentation and Test Procedure

All the beams were simply supported over the two steel plates and tested under four-point bending configuration. The uniformly increasing load was applied with an increment of 50 kN using a hydraulic actuator with a maximum capacity of 5000 kN. The surface of concrete was carefully rubbed with sand paper before bonding of CFRP sheets. To prepare two-part epoxy resin based on the manufacturer recommendations, the hardener and resin were mixed and stirred for three minutes with a proportion of one to four. After two-week curing of the beams, the CFRP sheets were installed and covered the shear span of beams at two sides.

The DEMEC discs were installed along and perpendicular to the strut centreline in all the beams with a distance of 200 mm equal to the length of DEMEC bar as shown in Figure 1. The resolution of DEMEC gauge was 0.001 mm. In an attempt to obtain the accurate average values of tensile strains for evaluating strut widening, the transverse tensile strains were measured in five positions with the same distance, perpendicular to the strut centreline as shown in Figure 1. The value of compressive strains was also measured along the strut centreline. The deep beams were tested at 28 days after casting.

EXPERIMENTAL RESULTS AND DISCUSSION

Prior to cracking, the elastic stress field occurred in the D-region of RC deep beams, which can be quantified with elastic analysis. Cracking occurred in the D-region as the applied load increased and disrupted the stress field which led to disorientation of the internal load path. The STM method rationally embodies a system of forces which is in equilibrium with a given set of loads. However, CFRP-strengthening disrupted the load trajectories in strut and complicated the disorientation of internal forces. Therefore, the estimation of CFRP-strengthened strut dimensions under loading became inaccurate.

Ultimate Shear Strength and Failure Mode of Beams

In this experiment, shear failure mode with diagonal crack propagating towards the load plate and support plate was dominant for unstrengthened RC deep beams as expected based on the prior research [12]. The application of steel cage under the loading plates and atop the support plates was successful as no bearing failure was observed in these areas. The CFRP-strengthening significantly increased the ultimate shear strength of RC deep beams as shown in Table 2. The rupture of CFRP sheets was dominant in shear failure of all CFRP-strengthened RC deep beams. Typical failure mode of unstrengthened and CFRP-strengthened RC struts in deep beams from the experiment is shown in Figure 4.

(a)

(b)

Figure 4: Typical failure of strut in RC deep beams with $a/d = 0.75$: (a) ordinary RC deep beam, (b) CFRP-strengthened RC deep beam.

Table 2: Ultimate shear strength of RC deep beams

a/d	$P_{u\text{-ordinary}}$ (kN)	$P_{u\text{-CFRP-strengthening}}$ (kN)
0.75	756.95	905.31
1.00	709.01	857.89
1.25	604.08	740.02
1.50	555.91	691.04
1.75	403.02	510.01
2.00	360.02	468.05

Strut Dimensioning

The orientation of strut is affected by the ratio of *a/d*. In addition, the behaviour of strut changes from elastic to plastic with the increase of applied load level. Therefore, the two variables of *a/d* ratio and applied load level were chosen for the evaluation. The experimental results showed that the strut deformation in the transverse direction was greater than the direction along the strut centreline for both the ordinary and the CFRP-strengthened specimens. This was because the strut tended to widen more than shorten due to the cracking occurring parallel to the strut centreline which caused greater values of transverse strains than compressive strains. Moreover, the effect of CFRP-strengthening on the strut compressive strain and consequently the strut shortening was insignificant.

Strut Widening

The cracking and crack widening caused softening behaviour in RC strut which affected strut dimensioning and consequently the strut behaviour under loading. The CFRP-strengthening of RC strut restricted the strut widening compared to the unstrengthened status.

General Investigation on RC Strut Widening: The load-transverse strain curves of the ordinary and the CFRP-strengthened RC struts from the experiment are shown in Figure 5. The figure demonstrates that the value of strut transverse strain for both the ordinary and the CFRP-strengthened RC struts increased with increasing of applied load. In addition, for the same applied load, the transverse strain and strut widening of the specimens with higher *a/d* were found to be greater than those with lower *a/d*. The difference between the transverse tensile strains of the ordinary and the CFRP-strengthened RC struts (□) with load step of 100 kN is presented in Table 3.

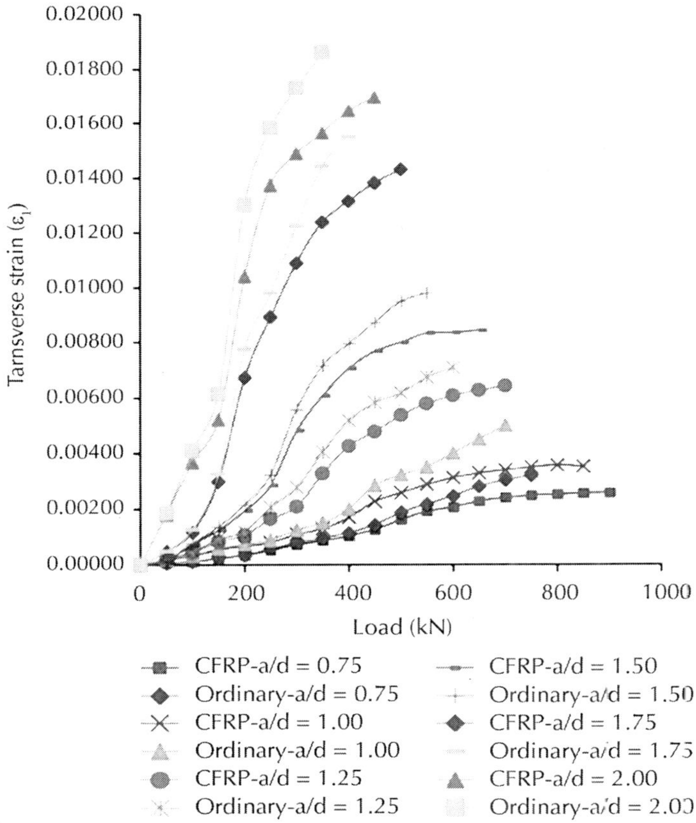

Figure 5: Load-transverse strain curves of unstrengthened and CFRP-strengthened RC struts.

Table 3: Difference of transverse tensile strains of the unstrengthened and CFRP-strengthened RC struts

	Load (kN)							
		100	200	300	400	500	600	700
a/d	0.75	0.0000	(0.0000)	0.0001	0.0001	0.0003	0.0004	0.0006
	1.00	0.0000	(0.0001)	0.0002	0.0003	0.0006	0.0009	0.0018
	1.25	0.0000	(0.0001)	0.0007	0.0009	0.0008	0.0010	—
	1.50	0.0000	(0.0002)	0.0008	0.0009	0.0015	—	—
	1.75	0.0000	(0.0010)	0.0013	0.0023	—	—	—
	2.00	0.0004	(0.0026)	0.0024	—	—	—	—

Based on Table 3, the increase of *a/d* ratio resulted in increase of □. Therefore, with the same applied load, the strut widening in the ordinary strut increased more than in the CFRP-strengthened strut as *a/d* ratio increased. It implied that the effect of CFRP-strengthening on restraining crack widening increased with the increase of *a/d* ratio. This was because the diagonal cracks were propagated in a larger area for high *a/d* ratio than those of low *a/d* ratio which resulted in the stress field distribution in a wider strut. Therefore, the stress field was transferred to different random portions of CFRP sheet through more diagonal cracks in high *a/d*beams compared to those of low *a/d*.

Based on Table 3, for the same *a/d* ratio, the increase of applied load resulted in greater value of □. Therefore, for the same *a/d*, the strut widening increased in the ordinary strut more than those of CFRP-strengthened strut as the applied load increased. This was due to the increasing crack widening in the high applied load compared to those of low applied loads. Thus, the effect of CFRP-strengthening appeared to be more apparent in higher applied load level.

Figure 5 illustrates the load-transverse strain curve of struts which consisted of two main parts for both groups of ordinary and CFRP-strengthened RC deep beams. The former was approximately a line extended up to about 40–60% of ultimate shear strength of beams. The linear trend of transverse strain-load curve was nearly similar for both ordinary and CFRP-strengthened struts. Thus, the effect of CFRP-strengthening on the strut widening was insignificant in this part of the curve due to the slight crack widening and CFRP sheet was not remarkably involved to restrain the crack widening. The latter was approximately a part of parabolic curve with extrinsic curvature as shown in Figure 5. The extrinsic curvatures part of the curves for CFRP-strengthened and ordinary RC struts became open and dissimilar from each other with increasing applied load level as opposed to the first linear part of curves. It was due to the fact that plastic behaviour of concrete was dominant in this part of curves instead of elastic behaviour which occurred in the linear part. Thus, the CFRP sheet was involved to restrain the crack widening, crushing of reinforced concrete and consequently transverse strains in CFRP-strengthened struts. Therefore, the amounts of transverse strains for ordinary strut remarkably grew faster than those of CFRP-strengthened struts, and it caused the parabolic parts of ordinary and CFRP-strengthened RC strut to go further away from each other.

Based on the experimental results, the value of transverse strain of ordinary RC struts remarkably grew faster than that of CFRP-strengthened RC struts after the amount of loading was higher than approximately 40–60% of ultimate shear strength of ordinary RC deep beams. Hence, the strut widening was remarkably affected by CFRP-strengthening after the applied load was higher than approximately 40–60% of ultimate shear strength of ordinary RC deep beams. The amount of □ was negligible in the linear part of curves which was due to minimal crack widening. Therefore, the effect of CFRP on the strut widening was insignificant in the low applied load level.

Strut Widening with Different Applied Load Levels: The average transverse strain of struts for both ordinary and CFRP-strengthened RC deep beams was classified in two groups according to applied load level. The strain-load curves were very close in distance and similar in pattern for the amounts of applied load lower than 50% of ultimate load as shown in Figure 5. Therefore, the applied load greater than 50% of ultimate load was taken into consideration in this section of the research as the strain-load curves were distinctly separated from each other in this level of loading.

The value of $\square_{50-75}$ represents the average of transverse strain corresponding to 50% and 75% of ultimate shear load. The same classification was used for 75% and 100% of ultimate shear load regarding the value of$\square_{75-100}$. The values of $\square_{50-75}$ and $\square_{75-100}$ for deep beams with *a*/*d* = 0.75 were assumed as a basis for normalisation as shown in Tables Tables44 and and5.5. The normalised values in Tables Tables44 and and55 depicted that, with the increase of *a*/*d*, the average transverse strain for ordinary and CFRP-strengthened struts rose to approximately 8 and 7 times compared to those of struts with *a*/*d* = 0.75, respectively. This increase followed approximately the same trend for both applied load classifications of 50–75% and 75–100% in ordinary RC deep beams. Based on Figure 6, the value of $B_{ordinary}$ grew faster than that of $A_{ordinary}$ as *a*/*d* increased. It implied that with the increase of *a*/*d* the strut widening and the increase rate of widening for load classification of 75–100% was greater than that of 50–75%. This was because the propagated cracks occurred in range of 50–75% applied load and then developed and widened when applied load reached the range of 75–100%. Thus, it resulted in greater widening of strut.

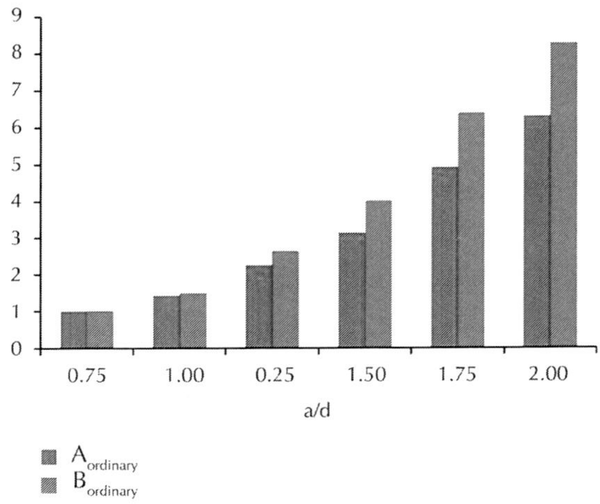

Figure 6: Comparison of average transverse strains of unstrengthened RC struts for different applied load levels.

Table 4: Average transverse strains of unstrengthened RC struts for different applied load levels

Number	*a/d*	50–75	75–100	$A_{ordinary}$	$B_{ordinary}$
1	0.75	0.0028	0.0032	1.00	1.00
2	1.00	0.0040	0.0048	1.43	1.50
3	1.25	0.0063	0.0084	2.25	2.63
4	1.50	0.0088	0.0128	3.14	4.00
5	1.75	0.0138	0.0204	4.93	6.38
6	2.00	0.0176	0.0264	6.29	8.25

Table 5: Average transverse strains of CFRP-strengthened RC struts for different applied load levels

No	*a/d*	50–75	75–100	A_{CFRP}	B_{CFRP}
1	0.75	0.0018	0.0025	1.00	1.00
2	1.00	0.0034	0.0040	1.89	1.60
3	1.25	0.0044	0.0058	2.42	2.30
4	1.50	0.0060	0.0080	3.33	3.20
5	1.75	0.0106	0.0143	5.89	5.72
6	2.00	0.0138	0.0187	7.67	7.48

The values of A_{CFRP} and B_{CFRP}, which were indicated in Figure 7, increased as *a/d* ratio increased. However, the rate of increase of B_{CFRP} was lower than that of A_{CFRP} as opposed to the ordinary strut as shown in Figure 7. Therefore, the increase rate of widening of CFRP-strengthened strut in load classification of 75–100% was lower than that of 50–75% as opposed to the ordinary struts. While beam was under the load classification of 50–75%, the CFRP sheet was stretched. Due to the restriction of crack widening by CFRP sheet, the strut widening was lower than the ordinary status. The aforesaid restriction by CFRP sheet was greater while beam was under 75–100% load classification because the CFRP sheet sustained higher tensile stress which was closer to its ultimate tensile stress compared to the ordinary status. The difference ofA_{CFRP} and B_{CFRP} became less apparent as *a/d* ratio increased as shown in Figure 7. This effect was due to the applied load for high values of *a/d* ratio which were proportionally less than low values of *a/d* ratio.

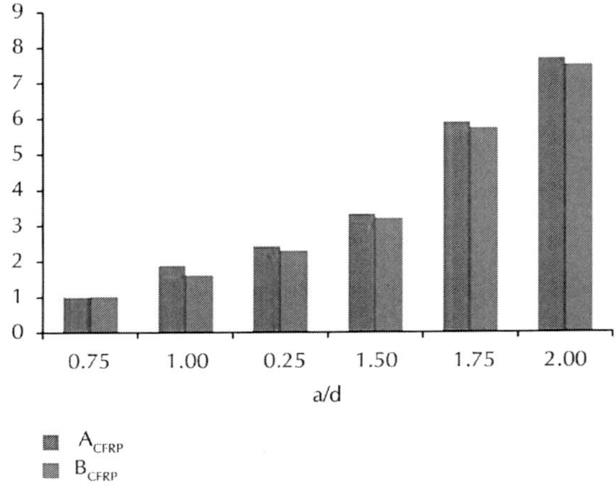

Figure 7: Comparison of average transverse strains of CFRP-strengthened RC struts for different applied load levels.

Shortening of RC Strut

The load-compressive strain curves consisted of two parts for both ordinary and CFRP-strengthened RC deep beams as shown in Figure 8. The former was a linear part extending from the beginning to the approximate amounts of compressive strains which was in the range of 0.0010 to 0.0015. The direction of linear part then turned downward sloping into a parabolic curve with an extrinsic curvature lower than the same curve for load-transverse strain. As opposed to the load-transverse strain, the curves of load-compressive strains for the ordinary and the CFRP-strengthened RC struts were close enough to appear almost similar which can be hardly differentiated as shown in Figure 8. However, the shortening amount for both of the ordinary and the CFRP-strengthened RC strut was not comparable with those of widening amounts. This was due to the deformation of the bottle-shaped strut which was mainly governed by the transverse stresses acting perpendicular to the strut centreline which significantly affected the widening of struts.

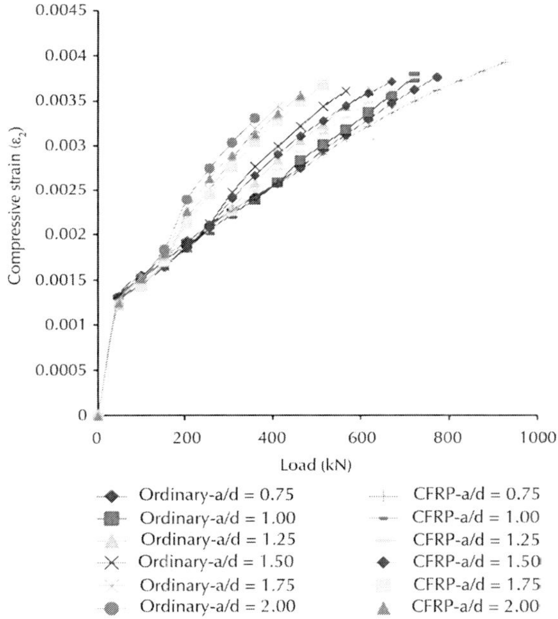

Figure 8: Load-compressive strain curve of unstrengthened and CFRP-strengthened RC struts.

The effect of CFRP-strengthening on the RC strut shortening was negligible. The curves in Figure 8indicated that the value of compressive strain for CFRP-strengthened RC strut was lower than that of the ordinary strut. CFRP sheet mechanical properties were conclusively not the cause of compressive strain reduction as this material is unable to sustain compressive stress. The increase of strut widening resulted in increase of strut shortening, and vice versa. The restraint of strut shortening was due to strut widening restrained by CFRP sheet. Therefore, the shortening of CFRP-strengthened RC struts was lower than that of ordinary struts.

CONCLUSIONS

Tests were performed on two sets of six RC deep beam specimens with and without CFRP-strengthening. The objective of the research was to investigate the RC strut deformation regarding its widening and shortening in the design of deep beams. The key variables were shear span to effective depth ratio and applied load level.

In deep beams subjected to load, as the level of widening of strut increases, the compressive strength of strut along its longitudinal axis decreases. Consequently the ultimate shear strength of deep beam decreases. Thus, designer of deep beam must pay attention to the significant effects of *a/d* ratio and applied load level on the widening of strut subjected to load. Nevertheless, the CFRP-strengthening of strut reduces the strut widening and increases the ultimate shear strength of deep beams. The experimental results regarding the effects of *a/d*ratio and applied load level on the strut widening for design purposes of deep beams with and without CFRP-strengthening are given below. The conclusion is drawn in terms of general aspects (items 1 and 2), similar widening behaviour (items 3, 4, and 9), distinctive widening behaviour (items 5, 6, 7, 8, 10, and 11), and shortening behaviour (item 12) of RC strut with and without CFRP-strengthening.

1. The CFRP-strengthening significantly increased the shear strength of RC deep beams.
2. Strut deformation in the transverse direction was greater than the direction along the strut centreline for both conditions of with and without CFRP-strengthening.
3. As the applied load increased the values of transverse strain in both the ordinary and the CFRP-strengthened RC struts increased.
4. With the same applied load, the transverse strain, and consequently the strut widening in both the ordinary and the CFRP-strengthened RC deep beams with high *a/d* was greater than those of low*a/d*.
5. With the same applied load, strut widening in ordinary strut increased more than CFRP-strengthened strut as the *a/d* increased.
6. The effect of CFRP-strengthening on crack widening restraint increased with the increase of *a/d*ratio. Thus, CFRP sheet contributed to strengthening of RC struts in high *a/d* ratio more

than in low*a/d* ratio.

7. With the same value of *a/d*, the strut widening of ordinary strut increased more than those of CFRP-strengthened strut as the applied load increased. Thus, the effect of CFRP-strengthening on restriction of strut widening appeared to be more apparent in higher applied load.
8. The value of transverse strain of ordinary RC struts remarkably grew faster than that of CFRP-strengthened RC struts after the amount of loading was higher than approximately 40–60% of ultimate shear strength of ordinary RC deep beams. Therefore, the strut widening was significantly affected by CFRP-strengthening after approximately 40–60% of ultimate shear strength of ordinary RC deep beams.
9. With the increase of *a/d* ratio up to 2.00, the average transverse strain of the ordinary and the CFRP-strengthened struts increased up to approximately 8 and 7 times compared to that of strut with *a/d* = 0.75, respectively.
10. With the increase of *a/d*, widening of ordinary RC strut increased, and the increase rate of widening for load classification of 75–100% was greater than that of 50–75%.
11. With the increase of *a/d*, widening of CFRP-strengthened RC strut increased. However, the increase rate of widening of CFRP-strengthened struts in load classification of 75–100% was lower than that of 50–75% as opposed to the ordinary struts.
12. The increase of strut widening resulted in the increase of strut shortening, and vice versa. The strut shortening was restrained as the CFRP sheet restrained the strut widening. Therefore, the shortening of CFRP-strengthened RC struts was lower than that of ordinary struts.

SUPPLEMENTARY MATERIAL

The Tables 1 and 2 represent the average value of transverse strain of ordinary and CFRP-strengthened RC struts from experiment respectively.

Table 1: Average transverse strain of ordinary unstrengthened RC struts

	El-ordinary					
Load (kN)-CFRP	a/d=0.75	a/d=1.00	a/d=1.25	a/d=1.50	a/d=1.75	a/d=2.00
0.00	---	---	---	---	---	---
50. 00	0.0001	0.0002	0.0002	0.0002	0.0005	0.0019
100.00	0.0001	0.0004	0.0005	0.0008	0.0013	0.0041
150.00	0.0003	0.0006	0.0009	0.0014	0.0033	0.0061
200.00	0.0004	0.0008	0.0012	0.0022	0.0077	0.0130
250.00	0.0006	0.0009	0.0021	0.0032	0.0098	0.0158
300.00	0.0008	0.0013	0.0028	0.0056	0.0123	0.0173
350.00	0.0010	0.0015	0.0040	0.0072	0.0144	0.0186
400.00	0.0012	0.0020	0.0052	0.0079	0.0155	---
450.00	0.0014	0.0029	0.0058	0.0087	---	---
500.00	0.0019	0.0032	0.0062	0.0095	---	---
550.00	0.0022	0.0035	0.0067	0.0098	---	---
600.00	0.0025	0.0040	0.0071	---	---	---
650.00	0.0028	0.0045	---	---	---	---
700.00	0.0030	0.0050	---	---	---	---
750.00	0.0032	---	---	---	---	---

Table 2: Average transverse strain of CFRP-strengthened RC struts

	£1-CFRP-strenghtnend					
Load (kN)-CFRP	**a/d=0.75**	**a/d=1.00**	**a/d=1.25**	**aid=1.50**	**a/d=1.75**	**a/d=2.00**
0.00	0.0000	0.0000	0.0000	0.0000	0.0000	0.0000
50.00	0.0001	0.0002	0.0002	0.0002	0.0005	0.0018
100.00	0.0001	0.0003	0.0004	0.0007	0.0012	0.0037
150.00	0.0002	0.0005	0.0008	0.0012	0.0030	0.0052
200.00	0.0003	0.0007	0.0011	0.0019	0.0067	0.0104
250.00	0.0005	0.0008	0.0016	0.0028	0.0089	0.0138
300.00	0.0007	0.0011	0.0021	0.0048	0.0109	0.0149
350.00	0.0009	0.0013	0.0033	0.0061	0.0124	0.0156
400.00	0.0010	0.0017	0.0043	0.0071	0.0132	0.0165
450.00	0.0013	0.0023	0.0048	0.0077	0.0138	0.0170
500.00	0.0016	0.0026	0.0054	0.0080	0.0143	

550.00	0.0019	0.0029	0.0058	0.0083	--	--
600.00	0.0021	0.0031	0.0061	0.0083	--	--
650.00	0.0023	0.0032	0.0063	0.0084	--	--
700.00	0.0024	0.0033	0.0064	--	--	--
750.00	0.0025	0.0033	--	--	--	--
800.00	0.0025	0.0035	--	--	--	--
850.00	0.0026	0.0035	--	--	--	--
900.0000	0.0026	--	--	--	--	--

ACKNOWLEDGMENTS

This research was supported by High Impact Research MoE Grant UM.C/625/1/HIR/MoE/ENG/54 from the Ministry of Education Malaysia. The authors acknowledge Dura Technology S/B for facilitating the experimental work.

SYMBOLS

- ε_{50-75}: The average of transverse strain corresponding to 50% and 75% of ultimate shear load
- ε_{75-100}: The average of transverse strain corresponding to 75% and 100% of ultimate shear load
- $\varepsilon_{ordinary}$: The average of transverse tensile strain of ordinary unstrengthened RC strut
- ε_{CFRP}: The average of transverse tensile strain of CFRP-strengthened RC strut
- $\Delta_{\varepsilon} = \varepsilon_{ordinary} - \varepsilon_{CFRP}$: The difference value between the transverse tensile strains of ordinary unstrengthened and CFRP-strengthened RC strut
- $A_{ordinary}$: The ratio of ε_{50-75} for each ordinary unstrengthened beam with a/d ranging in 1.00–2.00 toε_{50-75} for beam with $a/d = 0.75$
- $B_{ordinary}$: The ratio of ε_{75-100} for each ordinary unstrengthened beam with a/d ranging in 1.00–2.00 toε_{75-100} for beam with $a/d = 0.75$
- A_{CFRP}: The ratio of ε_{50-75} for each CFRP-strengthened beam with a/d ranging in 1.00–2.00 to ε_{50-75} for beam with $a/d = 0.75$
- B_{CFRP}: The ratio of ε_{75-100} for each CFRP-strengthened beam with a/d ranging in 1.00–2.00 to ε_{75-100} for beam with $a/d = 0.75$
- a/d: Shear span to effective depth ratio
- $P_{u\text{-}ordinary}$: Ultimate shear strength of ordinary unstrengthened RC deep beam from the experiment (kN)
- $P_{u\text{-}CFRP\text{-}strengthening}$: Ultimate shear strength of CFRP-strengthened RC deep beam from the experiment (kN).

CONFLICT OF INTERESTS

The authors declare that there is no conflict of interests regarding the publication of this paper.

REFERENCES

1. AASHTO. *LRFD, Bridge Design Specifications, Customary U.S. Units: 2008 Interim Revisions*. 4th edition. Washington, DC, USA: American Association of State Highway and Transportation Officials; 2008.
2. ACI. Building Code Requirements for Structural Concrete (ACI 318-08) and Commentary, section 10.7 and R10.7. 317-318, 2011.
3. *AS3600, Australian Standard for Concrete Structures*. North Sydney, Australia: Standard Association of Australia; 2009.
4. CAN/CSA-S6-06. Canadian highway bridge design code and S6.1-06 commentary on CAN/CSA-S6-06, Canadian Highway Bridge Design Code. Association Canadienne de Normalisation, 2006.
5. DIN. *Building and Civil Engineering Standards Committee. Plain, Reinforced and Prestressed Concrete Structures, Part 1: Design and Construction (DIN 1045-1)* Berlin, Germany: Deutsches Institut für Normung (DIN-Normen); 2001.
6. Eurocode2. EN 1992-2:2005, Design of concrete structures—Part 2: Concrete bridges—Design and detailing rules, 2005.
7. Wight JK, Macgregor JG. *Reinforced Concrete Mechanics and Design*. New Jersey, NJ, USA: Pearson Prentice Hall; 2009.
8. Hollaway LC. A review of the present and future utilisation of FRP composites in the civil infrastructure with reference to their important in-service properties. *Construction and Building Materials*.2010;24(12):2419–2445.
9. Panjehpour M, Farzadnia N, Anwar MP, Ali AAA. FRP sheet contribution in common repair techniques of concrete structures with emphasis on concrete columns. *International Journal of Sustainable Construction Engineering & Technology*. 2011;2(2):54–61.

10. Panjehpour M, Ali AAA, Anwar MP, Aznieta FN, Voo YL. An overview of Strut-and-Tie Model and its common challenges. *International Journal of Engineering Research in Africa*. 2012;8:37–45.

11. Panjehpour M, Ali AAA, Voo YL, Aznieta FN. Strut elaboration in strut-and-tie model. *Constructii: Journal of Civil Engineering Research*. 2012;2:45–53.

12. Kong FK. *Reinforced Concrete Deep Beams*. London, UK: Blackie; 1990.

13. Islam MR, Mansur MA, Maalej M. Shear strengthening of RC deep beams using externally bonded FRP systems. *Cement and Concrete Composites*. 2005;27(3):413–420.

14. El Maaddawy T, Sherif S. FRP composites for shear strengthening of reinforced concrete deep beams with openings. *Composite Structures*. 2009;89(1):60–69.

15. Lee HK, Cheong SH, Ha SK, Lee CG. Behavior and performance of RC T-section deep beams externally strengthened in shear with CFRP sheets. *Composite Structures*. 2011;93(2):911–922.

16. Nehdi M, Omeman Z, El-Chabib H. Optimal efficiency factor in strut-and-tie model for FRP-reinforced concrete short beams with (1.5 < *a/d* < 2.5) *Materials and Structures/Materiaux et Constructions*.2008;41(10):1713–1727.

17. Kim DJ, Lee J, Lee YH. Effectiveness factor of strut-and-tie model for concrete deep beams reinforced with FRP rebars. *Composites B: Engineering*. 2014;56:117–125.

18. Schlaich J, Weischede D. Detailing of concrete structures. *Bulletin d' Information: Comite Euro-International du Beton, Paris*. 1982;150:p. 163. (Ger).

19. Schlaich J, Schaefer K, Jennewein M. Toward a consistent design of structural concrete. *PCI Journal*.1987;32(3):74–150.

20. Rogowsky DM, Marti P. Bern, Switzerland: VSL International; 1991. Detailing for Post-Tensioning. (VSL Report Series).

21. Adebar P, Zhou Z. Bearing strength of compressive struts confined by plain concrete. *ACI Structural Journal*. 1993;90(5):534–541.

22. Panjehpour M, Ali AAA, Voo YL, Aznieta FN. Effective compressive strength of strut in CFRP-strengthened reinforced

concrete deep beams following ACI 318-11. *Computers and Concrete*.2014;13(1):135–165.

23. Panjehpour M, Ali AAA, Aznieta FN. Energy absorption of reinforced concrete deep beams strengthened with CFRP sheet. *Steel and Composite Structures*. 2014;16(5):481–489.
24. Panjehpour M, Ali AAA, Voo YL, Aznieta FN. Modification of strut effectiveness factor for reinforced concrete deep beams strengthened with CFRP laminates. *Materiale s de Construcción*. 2014;64(article 314)

Chapter 7

Additive Manufacturing of Al Alloys and Aluminium Matrix Composites (AMCs)

Diego Manfredi[1], Flaviana Calignano[1], Manickavasagam Krishnan[1,3], Riccardo Canali[1,2], Elisa Paola Ambrosio[1], Sara Biamino[2], Daniele Ugues[2], Matteo Pavese[2] and Paolo Fino[1,2]

[1] Center for Space Human Robotics @Polito, Istituto Italiano di Tecnologia, Corso Trento, Torino, Italy

[2] DISAT-Dipartimento Scienza Applicata e Tecnologia, Politecnico di Torino, Corso Duca degli Abruzzi, Torino, Italy

[3] DIGEP – Dipartimento di Ingegneria Gestionale e della Produzione, Politecnico di Torino, Corso Duca degli Abruzzi, Torino, Italy

INTRODUCTION

In a product development context, the term Rapid Prototyping (RP) was widely used to describe technologies which created physical prototypes directly from digital data. The first methods for rapid prototyping became available in the late 1980s and were used to produce models and prototype parts. Users of RP technology have come to realize that this term is inadequate to describe the more recent applications of these technologies. The ASTM F-42 committee was recently formed to standardize Additive Manufacturing (AM) terminology and develop industry standards. According to their first standard, ASTM F2792-10, AM is defined as "*The process of joining materials to make objects from 3D model data, usually layer upon layer, as opposed to subtractive manufacturing technologies*". The basic principle of this technology is that a geometric model, initially generated using three-dimensional Computer Aided Design (3D CAD) system (e.g. CATIA, Pro/Engineer, SolidWorks), can be manufactured directly without the need of process planning [1].

There are many related terms used to describe AM and common synonyms include: additive fabrication, additive layer manufacturing, direct digital manufacturing, 3D printing and freeform fabrication. Within the last 20 years, AM has evolved from simple 3D printers used for rapid prototyping in non-structural resins to sophisticated rapid manufacturing systems that can be used to create functional parts in different engineering materials directly without the use of tooling. Most work to date has been conducted using polymer materials, but the development of AM processes such as Selective Laser Sintering/Melting, Electron Beam Melting and Laser Engineered Net Shaping enabled to build parts by using metallic materials, metal matrix composites and ceramic materials. Additive manufactured parts are now utilized in aerospace, automotive, medical fields and also in consumer products and military [1-3].

Additive manufacturing or 3D printing is receiving unprecedented attention from the mainstream media, investment community, and national governments around the world. This attention reached a pinnacle when 3D printing was mentioned by United States President Barack Obama in his February 2013 State of the Union address. AM, just 25 years old and still a relatively small industry, has completed a transformation from obscurity to something that many can't stop talking about. In 2011, about 1600 articles were found. In 2012, they were 16000, a tenfold increase. Prototyping has been the technology's biggest application, thus the name rapid prototyping, and it remains a key category. The fastest-growing application, however, is in the actual manufacturing of parts for final products. In just 10 years, this important application has grown from almost nothing to more than 28% of the total global product and service revenues. The manufacturing of final parts, rather than prototyping, is where the manufacturing money is, and it is the most significant part of AM's future. Researchers and industry leaders in the European Union (EU) have identified AM as a key emerging technology. Teaming relationships have been formed between university, industry, and government entities within and across countries. The overall level of activity and infrastructure in the EU is greater to that of the U.S. in this key area. Several large cooperative projects have been funded, worth of millions of euros across Europe, among them AMAZE (Additive Manufacturing Aiming Towards Zero Waste & Efficient Production of High-Tech Metal Products) [4], RAPOLAC (Rapid Production of Large Aerospace Components) [5], Custom Fit (aiming at mass customized consumer and medical project manufacturing) [6], E-BREAK (Engine Breakthrough Components and Subsystems) [7], TiAlCharger (Titanium Aluminide Turbochargers-Improved Fuel Economy, Reduced Emissions) [8], and many others. Large aerospace companies, such as Boeing, GE Aviation, and Airbus, are hard at work qualifying AM processes and materials for flight. Boeing, for example, now has 200 different AM part numbers on 10 production platforms, including both military and commercial jets [9].

In 2013, NASA and Aerojet Rocketdyne have been testing rocket engine components made using additive manufacturing. At temperatures approaching 3300 °C, the AM rocket fuel injectors performed identically to conventionally manufactured parts. The series of tests demonstrated the ability to design, manufacture and test a critical rocket engine component using selective laser melting (SLM) manufacturing technology. This type of injector manufactured with traditional processes would take more than a year to make. but with these new processes it can be produced in less than four months, with a 70 percent reduction in cost [10]. The General Electric Aviation has used SLM to produce parts for its upcoming LEAP (Leading Edge Aviation Propulsion) family of turbofan engines, made in conjunction with France›s Snecma.

AM technologies have the potential to change the paradigm for manufacturing, away from mass production in large factories with dedicated tooling and with high costs, to a world of mass customization and distributed manufacture. AM can be used anywhere in the product life cycle from pre-production prototypes to full scale production, as well as for tooling applications or post production repair. AM processes are stimulating innovation in component design, enabling the manufacture of parts that cannot be made by traditional methods and are stimulating alternative business models and supply chain approaches. AM enables the manufacture of highly complex shapes with very few geometric limitations compared to traditional manufacturing processes.

This freedom of design, for example, has led to the technology being used to manufacture topologically optimized shapes with improved strength to weight ratios, a fundamental aspect in both aerospace and automotive design to reduce vehicle weight. For the bio-medical industry, AM is already leading to a revolution in customized medicine where dental implants, orthopedics, and hearing aids are manufactured to fit an individual's unique physiology.

However, despite many examples of growth and progress, many myths and misconceptions associated with the technology have developed:

- AM is a low-labor content "pushbutton" technology.

At the moment, a lot of work occurs before and after the actual production of the parts: in fact data needs to be prepared at the front end, which may require CAD expertise, the repair of the models, and optimization of support structures. Moreover, main building parameters, such as layer thickness, temperature, scan speed, and a number of other options may need to be adjusted for a particular part or type of material.

- AM is fast.

As the name implies, AM systems build parts by depositing, fusing, curing, or laminating consecutive layers of material. These layers are typically 0.025–0.250 mm in thickness, so parts often require thousands of layers. Therefore the preparation of large parts could take even several days.

- AM systems can produce anything.

Most of AM systems can successfully build shapes that cannot be fabricated easily or at all using conventional methods of manufacturing. However, AM processes also have limitations. One is minimum wall thickness. Another is the requirement for supports on down-facing surfaces, which can be difficult to remove. Material that is trapped in internal channels can also be difficult or impossible to remove, and the dimension of the internal channels determines the degree of difficulty in removing material.

- AM systems and materials are inexpensive.

Generally, industrial AM systems are more expensive than CNC machining centers, and starting materials are far more expensive (for example high quality powders, with controlled chemical composition, size and shape).

In any case, as AM continues to develop on multiple levels and in many directions, it will gain more momentum and respect as a method of manufacturing. Metal parts from some AM systems are already on par with their cast or wrought counterparts. As organizations qualify and certify these and other materials and processes, the industry will grow very large. 3D printing is not competing with conventional manufacturing techniques, but is instead complementing and hybridizing with them to make new things possible.

ADDITIVE MANUFACTURING PROCESSES FOR METALS

It was less than 10 years ago that these systems began to gain traction commercially. Today, seven companies, all in Europe, offer systems based on powder bed fusion technology. Examples of powder bed fusion are direct metal laser sintering (DMLS) from EOS and selective laser melting (SLM) from Renishaw. The most popular metal systems use a laser to heat and melt fine particles in a powder bed. Parts obtained with these systems are impressive, especially the fine features and small internal channels that are possible.

Systems that instead use an electron beam as the energy source are much faster, but the surface finish and feature detail are not as impressive. In the case of some orthopedic implants, however, a rough surface is preferred. This, coupled with speed, is why many medical implant manufacturers have selected electron beam melting (EBM) systems from Arcam (Arcam is currently the only manufacturer to use electron beam energy in a powder bed). Many metal parts that are currently being made by AM can be cost prohibitive, difficult, or impossible to produce using conventional methods, such as casting or machining.

This gives AM an advantage, but only if the right types of parts are selected. Quality requirements play a role. Good feature detail and surface finish are possible and comparable to metal castings, but they do not match the surface quality of CNC machined parts. If a part can be produced conventionally at a reasonable cost and the volume is relatively high, it is often best to go that route at the present time. The metal AM systems are complemented by extensive CNC machining and other traditional processes.

Many types of metals are available on metal AM systems. Among the most popular for medical and aerospace applications is the titanium alloy Ti-6Al-4V. Other metals used are cobalt–chrome, stainless steels, tool steels, aluminium alloys such as AlSi10Mg, jewelry and dental gold alloys, nickel-based superalloys such as Inconel 625 and 718, and TiAl alloys. Aerospace-grade aluminium and other metals are in development. All of the unused metal powder can be recycled in the machines after sieving. Industry standards are becoming increasingly important as companies apply AM to the production of final products. In January 2012, ASTM International Committee F42 on Additive ManufacturingTechnologies approved F2924-12 Standard Specification for Additive Manufacturing Titanium-6Aluminium-4Vanadium with Powder Bed Fusion. It is the first AM material standard by ASTM and it could give a boost to the additive manufacture of Ti-6Al-4V.

The technology classification agreed by the AM SIG (Special Interest Group) can be seen detailed inTable 1, which shows seven top level classifications for additive manufacturing technologies, below which there are a range of different material classifications, and discrete manufacturing technologies produced by a range of global companies.

TABLE 1: Classification of additive manufacturing processes defined by the AM SIG – adapted from ASTM AM classifications (2012) [11]

Classification	**Technology**	**Description**	**Materials**	**Developers (Country)**
Binder Jetting	3D Printing Ink-jetting S-Print M-Print	Creates objects by depositing a binding agent to join powdered material	Metal, Polymer, Ceramic	ExOne (US) VoxelJet (Germany); 3D Systems (US)
Direct Energy Deposition	Direct Metal Deposition Laser Deposition Laser Consolidation Electron Beam Direct Melting	Builds parts by using focused thermal energy to fuse materials as they are deposited on a substrate	Metal: powder and wire	DM3D (US) NRC-IMI (Canada) Irepa Laser (France) Trumpf (Germany) Sciaky (US)
Material Extrusion	Fused Deposition Modeling	Creates objects by dispensing material through a nozzle to build layers	Polymer	Stratasys (US) Delta micro Factory (China) 3D Systems (US)
Material Jetting	Polyject Ink-jetting Thermojet	Builds parts by depositing small droplets of build material, which are then cured by exposure to light	Photopolymer, Wax	Stratasys (US) LUXeXcel (Netherlands) 3D Systems (US)
Powder Bed Fusion	Direct Metal Laser Sintering Selective Laser Melting Electron Beam Melting Selection Laser Sintering	Creates objects by using thermal energy to fuse regions of a powder bed	Metal, Polymer, Ceramic	EOS (Germany) Renishaw (UK) Phenix Systems (France) Matsuura Machinery (Japan) ARCAM (Sweden) 3D Systems (US)

Sheet Lamination	Ultrasonic Consolidation Laminated Object Manufacture	Builds parts by trimming sheets of material and binding them together in layers	Hybrids, Metallic, Ceramic	Fabrisonic (US) CAM-LEM (US)
VAT Photopolymerisation	Stereolithography Digital Light Processing	Builds parts by using light to selectively cure layers of material in a vat of photopolymer	Photopolymer, Ceramic	3D Systems (US) EnvisionTEC (Germany) DWS Srl (Italy) Lithoz (Austria)

DIRECT METAL LASER SINTERING — DMLS

Direct Metal Laser Sintering (DMLS), a trademark of EOS GmbH (Germany), is an Additive Manufacturing (AM) technology that creates parts in a layer-by-layer fashion directly from computer-aided design by selectively fusing and consolidating thin layers of powders with a scanning laser beam. DMLS is a net-shape process, producing parts with high accuracy and detail resolution, good surface quality and excellent mechanical properties. As mentioned before, it has many benefits over traditional techniques and its application for manufacturing three dimensional objects represents one of the promising directions to solve challenging industrial problems. Moreover, since the components are built layer by layer, it is possible to design internal features and passages that could not be cast or otherwise machined. Complex geometries and assemblies with multiple components can be simplified to fewer parts with a more cost effective assembly. DMLS does not require special tooling like castings, so it is convenient for short production runs. Applications using this technology include direct parts for a variety of industries including automotive and other industries that use complex parts of small to medium size. For instance, concerning space applications, when the "buy to fly ratio" of a structure is below about 30%, meaning that more than 70% is machined away, additive manufacturing technologies become an alternative to conventional machining for complex metallic parts. And when direct manufacturing allows to produce in one shot spacecraft structures resulting from complex assemblies, then the offered possibilities are really huge. DMLS has also already been used to fabricate lightweight structures, similar to cellular structures. Inspiring to natural systems, a variety of synthetic cellular solids could be made including stochastic foams, consolidated powders, hollow sphere structures, honeycombs, textile laminates and lattice block or miniature truss structures. The cellular metal structures have been used in various industrial applications such as heat exchangers, in automotive and aerospace industries, thanks to their valuable characteristics as low density, high strength, good energy absorption and good thermal properties. Virtually any material can be produced by additive manufacturing, most of those by DMLS, with the exception of some intermetallic and high temperature alloys.

One of the most interesting types of these materials are aluminium alloys. It is rather easy to produce alloys with compositions suitable for casting, since they will melt easily, but recently wrought alloys have also been fabricated by this technique, and new ones are being studied. The microstructure of these alloys is peculiar, since the very fast cooling occurring after the melting induced by laser provides an ultra-fine microstructure and this is responsible for a significant increase of the mechanical properties that can be obtained with conventional casting processes followed by heat treatments. It is also possible, to increase specific properties of the alloy, to produce aluminium matrix composites (AMC). These could have applications in high-demanding components, for instance in automotive and aerospace industries, where it is necessary to improve stiffness, hardness and high temperature properties. Ceramic discontinuous reinforcements are the most suitable reinforcement that can be envisaged.

DMLS PROCESS

Initially, all contour of the layer structure is exposed with a selected laser power and contour speed. As the diameter of the melted zone is usually larger than the laser diameter, it is necessary to compensate the dimensional error and the laser beam must be shifted by half the width from the contour to the inside, to make sure that the contour of the later part will correspond exactly to the original CAD data. This correction of the position is called Beam Offset (BO) [12]. The BO value is again defined with respect to the edge of the boundary (Figure 1), and if this value is higher or less than the correct value, the particles of the irradiated region may be not melted or over-melted. During hatching, the laser beam moves line after line several times to assure that the melting process can unroll completely. The distance between the lines is called hatching distance (*hd*) and is set about one quarter of the laser beam.

Another important parameter that can lead to a distorted part or a process interruption is the layer thickness. If the value is too high, no optimal adhesion between the single layers can be realized because the melting depth is not high enough. Furthermore, mechanical tension can be generated through this layer which can lead to detachment of the layer below. If the selected value is too small, a tearing-off of a structure can happen during the recoating process, since the melted particles get struck between it and recoater blade [12].

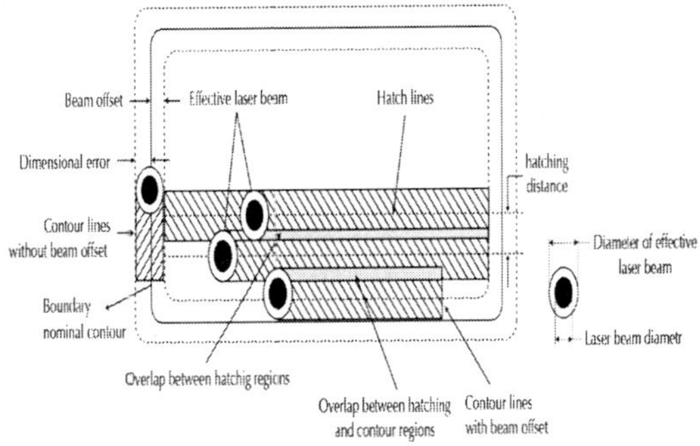

Figure 1: Exposure strategies.

DMLS SCANNING STRATEGY

Many scanning options are provided in commercial SLM machines which include skywriting and hatch pattern along *x*, *y*, both in *xy* and alternating in *xy* for different requirements. DMLS EOS machine has up-skin and down-skin options to improve mechanical properties by allowing user to assign different process parameters at adjoining layers.

Subsequent paragraphs illustrate salient features of scanning strategies available with DMLS process.

SKYWRITING

During scanning a certain time is needed to accelerate the mirrors to the desired speed. This is due to inertia of mirrors used for scanning. During this time laser beam cover some distance in which speed is not constant hence more energy is applied at the edges of the part than the inside of the part. To avoid this situation and to keep acceleration and retardation phase out of the layer geometry, skywriting option is provided in DMLS machines. If skywriting option is selected, then the mirror is accelerated already before the start of the part so that it has reached the desired speed before the beginning of exposure. Laser is switched on at the start of the part. Similarly retardation phase begins at the part end where the laser is switched off as shown in Figure 2.

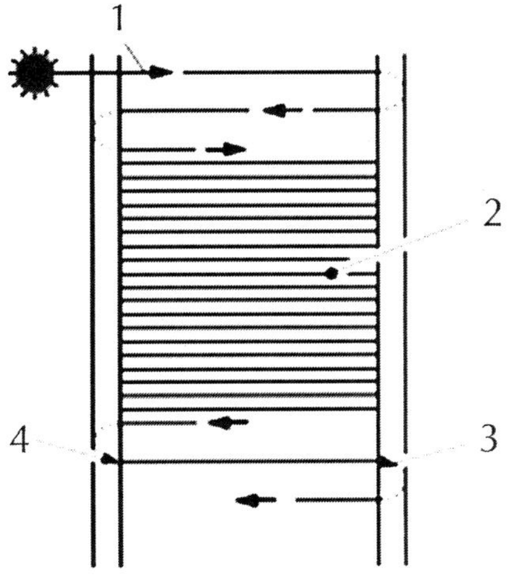

Figure 2: The exposure vector is shown by a solid line (1). The dashed line presents the skywriting. In this area the laser beam is run with a laser power of 0 W (3). With the beginning of the part (4), the laser power switched on with a sized value and the laser beam is running with a constant velocity.

HATCH PATTERN

Four choices for hatch pattern selection are generally available, i.e., along *x*, along *y*, both in *xy* or alternating in *xy* as shown in Figure 3. Scanning can be done either along *x* or along *y* (Figure 3a and 3b). If both in *x and y* options are selected than there will be double exposure on the layer, once along*x* and then along *y* (Figure 3c). In alternating in *xy* choice, direction of scanning is changed for alternating layers (Figure 3d). Figure 3e shows the direction of scanning rotated of 67° between consecutive layers. This is default value of the hatch pattern in DMLS EOS machine.

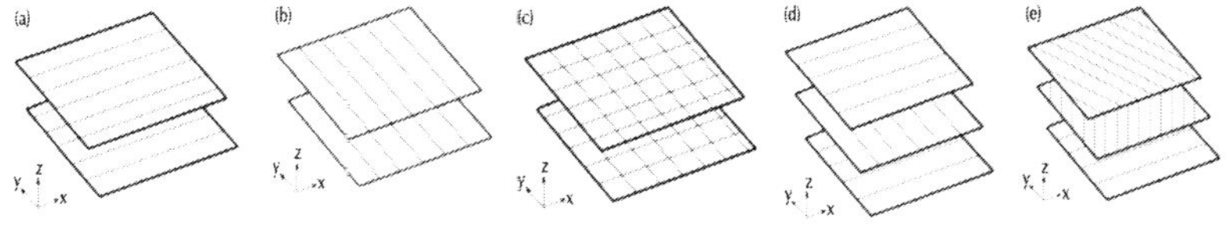

Figure 3: Different hatch patterns or scanning strategies.

UP-SKIN AND DOWN-SKIN

The layer above which there is no area to be exposed, is termed up-skin, and is built in three layers. Similarly the layer below which there is no exposed area is called down-skin, and is built in two layers. The rest of the part is termed core or in-skin, and is shown in Figure 4. Commercial machine software calculates up-skin, down-skin and core areas and different parameters can be assigned for these areas. Some overlapping area between core and up-down-skin area can also be assigned for improved joining among these zones.

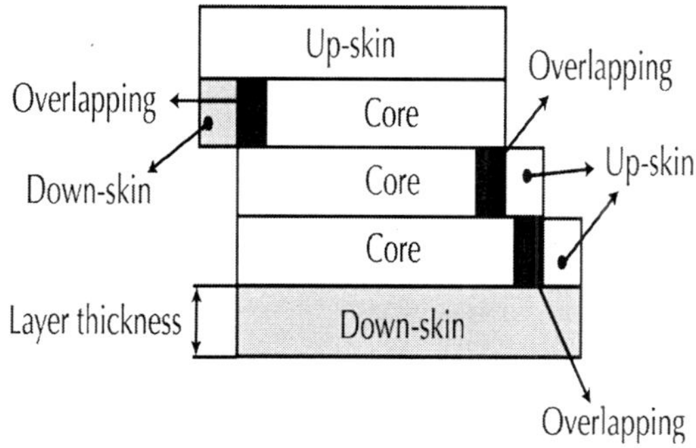

Figure 4: Up-skin, down-skin and core regions.

MATERIALS PROCESSED BY DMLS

Table 2 condenses a list of metal-based materials processed by SLM/ DMLS as available in the literature.

Table 2: Examples of materials used in SLM/DMLS

Family	**Material**	**References**
Metals and alloys	Steel	hot-work steel [13]
		stainless steel 316L [14-17]
		martensitic steel [18]
		tool steel [15-19]
		maraging steel [20]
	Titanium	Ti6Al4V [21-24]
		Ti6Al7Nb [25]
	Nickel based alloy	Inconel 718 [26]
		Inconel 625 [27]
	Copper	copper [28-29]
	Gold	Gold [30]
	Aluminium	A6061 [31]
		AlSi12 and AlSi12Mg [32,33]
		AlSi10Mg [34-37]

Composites	MMC	Fe-graphite, Ti- graphite/diamond, Ti-SiC, AlSi-SiC, AlMg-SiC, Co-WC, Fe-SiC and Cu, Ni, Ti, C, Cu-TiC and Cu, Ni, Ti, B_4C, Cu-TiB_2[38]
		Al4.5Cu3Mg-SiC [39]
	CMC	ZrO2, Y_2O_3, Al, Al_2O_3 and TiO_2,Al,C TiC/Al_2O_3 [38]

Some of the main common features for different materials processed by SLM are summarized hereafter.

***Microstructure*:** small grains, non-equilibrium phases and new chemical compounds are formed during SLM of metal powders [28,40].

Mechanical properties**:** due to the formation of very fine grains and non-equilibrium structures, the SLM parts could exhibit better mechanical properties (yield strength, ultimate tensile strength and ductility) than the wrought products [28]. Yasa et al. [20] have studied the Charpy impact test of TiAl6V4, A316L stainless steel and maraging steel parts produced by SLM. Spierings et al. [41] have investigated the influence of three different size distribution of 316L stainless steel powder on mechanical properties and surface quality. Their research concluded that powder with smaller size particle distribution could be easily melted and yields high density, high mechanical strength and productivity.

Anisotropy: the parts are produced by stacking layers in SLM, due to this reason the parts have some anisotropy in a particular direction. Buchbinder et al. [42] have investigated the fatigue testing of AlSi10Mg parts produced by SLM. Their investigation concluded that sample produced in x plane exhibited higher fatigue resistance than samples produced in other orientations. Yadroitsev et al. [27] have studied the mechanical properties of SLM parts. They found that Young's modulus of Inconel 625 parts produced in xy plane are 1.5 greater than that of parts produced in z direction. Qiu et al. [43] studied the microstructure and tensile properties of Ti6Al4V parts produced by SLM in two different orientations. They found that Ti6Al4V parts show high tensile strength, but the Ti6Al4V samples oriented along z axis showed higher ductility than that of sample oriented in xy plane. This anisotropy was caused by the orientation of the columnar grains with respect to the tensile test direction. Thijs et al. [44] studied about SLM of tantalum parts. Their study revealed that grains were oriented in <111> direction, because of partial re-melting of the previous layer, competitive growth mechanism and specific global direction of heat flow during SLM of tantalum. This texture formation had a large influence on yield strength during compression test of tantalum parts.

Residual stresses: these are induced in the SLM parts due to the complexity in melting behavior of layers. Re-melting and solidifying of previously melted layers can cause stress field to exceed the yield strength of the material, causing distortion of the part. Studies show that island scan strategy could reduce the residual stress of SLM parts [45]. Shiomi et al. [46] measured residual stress of chromium molybdenum steel parts produced by SLM. They found that: performing stress relieving treatment of build platform along with parts at about 700 °C for one hour, decreased residual stress by 70%, re-scanning of each layer reduced the residual stresses by 55% and heating of build platform to 160 °C reduced the residual stresses by 40%. Mercelis and Kruth have used crack compliance method and XRD to measure residual stress of 316L stainless steel parts produced by SLM. Zaeh and Branner [47] have carried out Finite Element Analysis to evaluate residual stress of steels processed by SLM and they also measured residual stress by using neutron diffraction.

Surface Roughness: the surface roughness of the SLM parts is higher than that of conventionally produced parts. Vertical faces of the part have lower surface roughness than that of curved or inclined or top faces. To improve the surface finish of the SLM parts, secondary operations like shot peening, machining or abrasive jet machining are carried out. Kruth et al. [48] showed that the surface quality of Stainless steel 316L and Ti6Al4V parts could also be improved in SLM by using Selective Laser Erosion (SLE) and re-melting. Calignano et al. [12] have improved the roughness of AlSi10Mg parts produced by DMLS process.

CASE STUDY — ALSI10MG PARTS BY DMLS

Experimental Equipment

Aluminium alloy samples were produced by DMLS with an EOSINT M270 Xtended version. In this machine a powerful Yb (Ytterbium) fiber laser system in an argon atmosphere is used to melt powders with a continuous power up to 200 W and a spot size of 100 μm. The aluminium powder alloy used in this study is a gas atomized one produced by EOS Gmbh (Germany), and its nominal composition is reported in Table 3.

Table 3: Nominal composition of EOS AlSi10Mg alloy powder in accordance with standard DIN EN 1706:2010-06

Element	Si	Fe	Cu	Mn	Mg	Zn	Ti	Al
Weight %	9-11	≤0.55	≤0.05	≤0.45	0.2-0.45	≤0.1	≤0.15	remainder

This alloy is similar to an A360.2 alloy [49]: thanks to its near eutectic composition in the Al-Si phase diagram it is often used in casting, having a melting temperature around 570 °C. This alloy offers good strength and hardness and is therefore used for parts with thin walls and complex geometry subjected to high loads, as in the aerospace and automotive industries.

Table 4 shows the default values assigned to up-skin, down-skin and core for the scan speed, laser power and hatching distance parameters in producing the Al alloy specimens.

Table 4: Default values of scan speed, laser power and hatching distance for EOS M270 Xtended

Parameters	**Core**	**Up-skin**	**Down-skin**	**Contour**
Scan speed (mm/s)	800	1000	900	900
Laser power (W)	195	195	190	80
Hatching distance (mm)	0.17	0.2	0.1	

Samples of rectangular shape and 50 × 10 × 3 mm size were produced to analyze the density, hardness and Young's Modulus. Considering tensile tests, specimens were built according to the standard ASTM E8M. For the analysis on the surface roughness, the three input or control factors chosen for up-skin were scan speed, laser power and hatching distance (Table 5). The surface finish of the samples was then analyzed though a 3D scanner ATOS Compact Scan 2M (GOM GmbH) and a MarSurf M 300C (Mahr GmbH) mobile roughness measuring instrument before and after post-processing. To compare the different DMLS samples, the average value of the ordinates from centerline, defined as R_a, was used. It is theoretically derived as the arithmetic average value of departure of the profile from the mean line along a sampling length. The shot-peening process was performed with glass microspheres using a sand-blasting machine, SD9 Northblast, in order to improve their surface finishing. The samples, before and after shot peening, were also characterized by a Field Emission Scanning Electron Microscope (FESEM) Zeiss SupraTM 40.

TABLE 5: Process parameters values for the up-skin

Variable Parameters	**Values**
Scan speed (mm/s)	800, 850, 900, 950, 1000, 1250
Laser power (W)	120, 155, 190
Hatching distance (mm)	0.10, 0.15, 0.20
Fixed Parameters	**Values**
Layer thickness (μm)	30
Spot size (mm)	0.1

Powder Analysis

The AlSi10Mg powder was gas atomized and supplied by EOS GmbH. A preliminary observation was made to investigate the morphology of the AlSi10Mg powder particles by using the FESEM described above.

The spherical morphology and smooth surface of the powder particles result in a good flowability and homogeneous layer distribution [36]. Figure 5 shows the FESEM micrograph of AlSi10Mg powder in as received condition. The size of AlSi10Mg powder ranges from 1 to 35 μm with an average size around 23 μm. Bigger clusters of dimension larger than 40 μm were also observed. Thus the powder was sieved to separate the agglomerated powder particles each time before producing the final part in DMLS process.

Figure 5: FESEM observation of AlSi10Mg powder.

In Table 6 are summarized the mean values for density, Young's modulus and Vickers Hardness. Moreover, considering for the AlSiMg alloy a theoretical density of 2.68 g/cm³ [49], the percentage of the residual porosity can be calculated.

Table 6: Density, Hardness and Young's Modulus of aluminium alloy DMLS specimens.

Density (g/cm³)	**Residual Porosity (%)**	**Hardness (HV)**	**Young's Modulus E (GPa)**
2.66	0.8	105 ± 5	73 ± 1

Considering tensile tests, the results are summarized in Table 7. Variations were not found among performances of samples with different orientations on the powder deposition plane, while there are some differences with the values obtained along the direction perpendicular to it, as already discussed in a previous paragraph. In fact, as the name implies, the parts are produced by stacking layers in SLM, due to this reason the parts have some anisotropy in the so called "building direction", or z axis.

Table 7: Mean values of tensile properties of aluminium alloy DMLS specimens produced according the standard ASTM E8M, compared to a similar alloy in as-fabricated conditions [*]

Material	**Orientation**	**Yield Strength [MPa]**	**Ultimate Tensile Strength [MPa]**	**Elongation at break [%]**
AlSi10Mg after DMLS	xy-plane	240 ± 8	330 ± 4	6.2 ± 0.4
	z axis	230 ± 5	328 ± 4	4.1 ± 0.3
A360.0 F *	–	170	317	5

The AlSi10Mg DMLS specimen microstructure was analyzed by optical microscopy (Leica DMI 5000 M optical microscope) and by FESEM. Top and a lateral surfaces of the specimens were polished by using SiC abrasive papers of different grits and then by using 6 μm, 3 μm and 1 μm diamond pastes respectively. Then they were etched with Weck's reagent or HF in water to highlight the scanning tracks or melt pools. The dimensions of the melt pools (width and depth) depend on the laser power and hatching distance employed, and also on the scanning strategy (Figure 3).

Figure 6 shows the optical microscopy images of AlSi10Mg specimen. Hauser et al. have reported that the cross-section of the single tracks formed in SLM process were either crescent shape or elliptical section [50]. From the optical micrograph of the AlSi10Mg sample polished in the lateral surface (z direction), a typical crescent shape structure can be noticed, if the scan track or melt pool is oriented perpendicular to the polished surface (Figure 6, on the left). Some elongated scan tracks are also observed, when the scan track is aligned at an angle with respect to the polished surface. This is cue to the scanning strategy of the laser beam adopted (see Figure 3e). Figure 6 shows also the optical microscopy image of AlSi10Mg specimen polished in xy plane. The cross section of scan tracks of different layers and the overlapping of scan tracks can be observed.

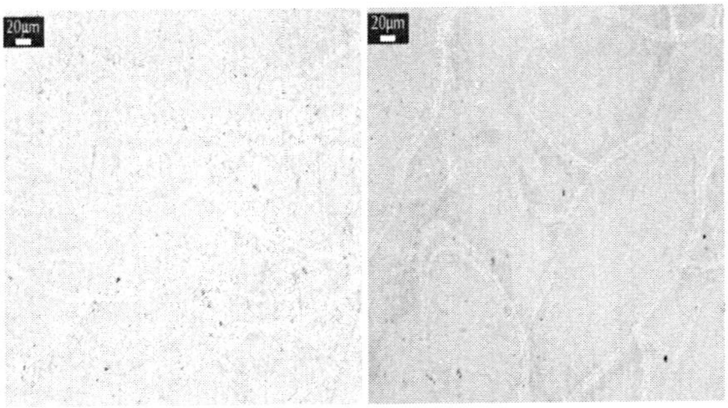

Figure 6: Microstructure of AlSi10Mg along *Z* direction (on the left) and along *XY* plane (on the right) after etching with HF in water reagent.

The geometry of the scan tracks was not clear with the default scanning strategy, because the scan lines are always rotated by 67° with respect to the scan lines of previous layer (see Figure 3e). In order to observe the geometry of the scan tracks, a specimen was produced by orienting the scan lines alongx axis. Figure 7 (a) and (b) show the microstructure of AlSi10Mg sample produced by orienting scan tracks only along *x* axis (see Figure 3b).

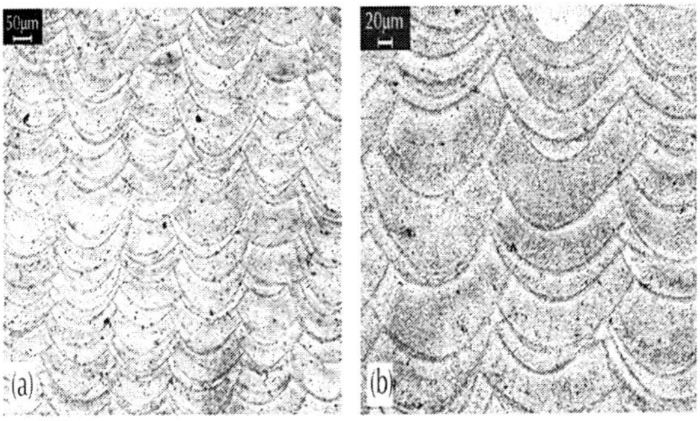

Figure 7: Optical micrographs of AlSi10Mg specimen produced by orienting scan tracks along x axis, after etching with Weck's reagent [51].

It can be observed that the shape of scan tracks is crescent or elliptical. However, this specimen had a higher porosity with respect to the ones obtained with the default scanning strategy. Therefore, the default scanning strategy with 67° is generally chosen to produce AlSi10Mg samples for the measurement of mechanical properties.

To appreciate the small grain size, the section along the build direction was observed by FESEM, focusing on a region between adjacent melt pools, as shown in Figure 8. After etching with HF and going at higher magnification, it can be seen that the melt pool contour region is characterized by a fine cellular-dendritic structure made by agglomerates of grains with mean diameters of a hundred of nanometers or less [37]. It was found that these agglomerates are different in length, thickness and aspect ratio, probably due to the different thermal heat fluxes during each scanning track.

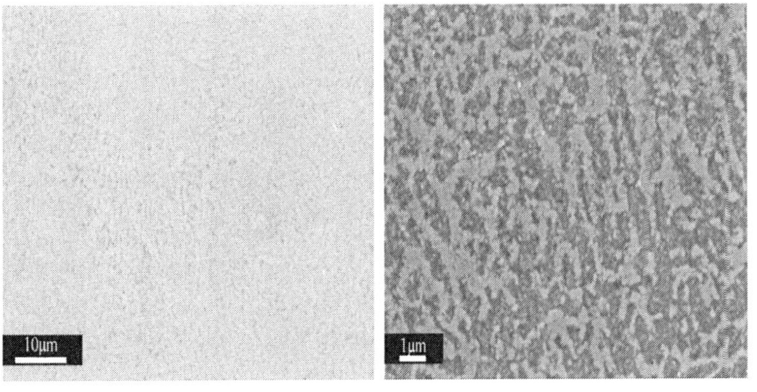

Figure 8: FESEM observations of AlSi10Mg specimen before etching (on the left), and after etching with HF (on the right) at higher magnification.

Also, fracture surfaces after tensile tests were investigated by FESEM, as reported in Figure 9. As can be seen, the surface is covered by very fine dimples, clearly visible only at high magnification. On the left image there can also be observed two little concave zones probably related to two spherical particles not completely melted. At higher magnification is possible to appreciate the very fine dimension of the microstructure, with presence of particles of tens of nanometers.

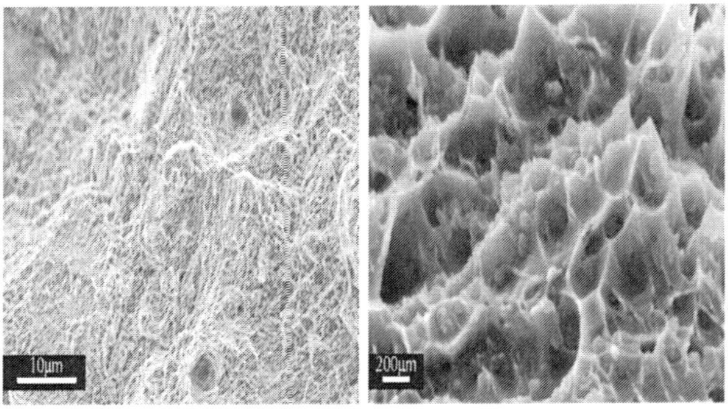

Figure 9: FESEM images of an aluminium alloy DMLS fracture surface covered by sub-micrometric voids and dimples with a nanometric size.

Effect of Heat Treatments

Due to the high thermal gradients, DMLS process fabricates parts with high residual thermal stresses. In order to avoid the bending of the AlSiMg parts fabricated, before removing them from the building platform, it is fundamental to perform an annealing for 2 h at 300 °C. Annealing offers very good dimensional and physical stability of the parts. It also provides the lowest level of residual stresses. The cooling is done in furnace or in the still air.

Apart from this, it is not common to perform a post heat treatment such as for conventional casting alloys. However, it is well known that mechanical properties of aluminium alloys are affected by heat treatments. Hence, the influence of heat treatments on hardness and tensile properties of AlSi10Mg parts produced by DMLS process was also investigated. In particular, annealing (T2), T4 and T6 heat treatments were carried out on fifteen AlSi10Mg samples for mechanical testing instead of stress relieving [51]. After the heat treatment, the hardness and tensile properties of these specimens were evaluated.

The first specimens were subjected to annealing treatment at 530 °C for 5 hours and cooled in the furnace. The second samples were obtained in T4 heat treatment: solution heat treated at the temperature of 530 °C for 5 hours and then quenched in water, followed by a room temperature ageing of at least two weeks. Finally the last samples were subjected to a T6 heat treatment cycle [49]. In this case, the AlSi10Mg specimens were solution heat treated at the temperature of 530 °C for 5 hours, then quenched in water and artificially aged at 160 °C for 12 hours.

The specimens were polished and the micro hardness was measured by using a Leitz instrument. The indentation load used was 50 g and duration of the indentation was 30 seconds. The results are illustrated in Table 8. The hardness after the T6 heat treatment is 13% higher than the mean value of the laser sintered AlSi10Mg samples just after the stress relieving treatment (105 HV, Table 6). Considering the T4 heat treatment, it reduced hardness by 15 % compared to AlSi10Mg sample without heat treatment, while with annealing the mean hardness was reduced by 58 %.

Table 8: Effect of heat treatment on hardness of AlSi10Mg samples

Type of heat treatment	Hardness (HV)	
	mean	S.D.
Annealing	46	3
T4	89	7
T6	119	6

Tensile tests were also performed on AlSi10Mg after heat treatments. Again the specimens were produced according to ASTM E8M. The specimens were produced in the xy plane orientation by using DMLS process. Post processing operations such as shot peening or machining was not carried out on these specimens after heat treatments. The tensile tests of heat treated AlSi10Mg specimens were carried out by using EasyDur 3MZ-5000.

The results of the tensile tests are summarized in Table 9 [51].

Table 9: Tensile properties of AlSi10Mg specimens after heat treatments [51]

Type of heat treatment	Rp0.2 (MPa)		UTS (MPa)		%A	
	mean	S.D	Mean	S.D	Mean	S.D
Annealing	72	7	113	3	12.6	0.9
T4	131	9	227	4	6.9	0.8
T6	245	8	278	2	3.6	0.8

It can be concluded that all the heat treatments reduced the tensile properties of AlSi10Mg specimens (see also Table 7). Annealing reduced the ultimate tensile strength by 66 %, when comparing to the tensile strength of AlSi10Mg specimens in "as built" condition (just after DMLS, without shot peening). However, the elongation was improved by 103%, as expected. The tensile strength of the AlSi10Mg specimens after T4 treatment was reduced by 31% and ductility was improved by 11%, respectively. The ultimate tensile strength and elongation of the AlSi10Mg samples after T6 heat treatment were decreased by 16% and 42%, respectively. The only improvement observed after heat treatment is a slight improvement in yield strength, from 240 to 245 MPa, for the T6 samples with respect to the "as built" condition.

Stress-strain curves obtained from tensile tests of specimens after different heat treatment conditions are compared in Figure 10.

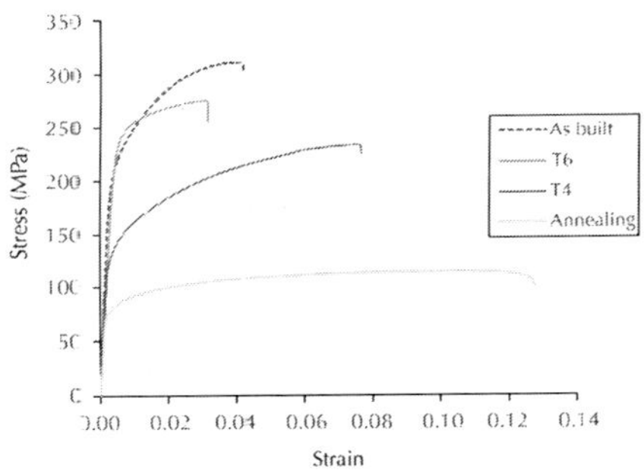

Figure 10: Stress vs strain curves of AlSi10Mg specimens in "as built" condition and after heat treatments [51].

FESEM analysis of the fracture surfaces after different heat treatments were also performed, and the micrographs are shown in Figure 11. It can be seen that the dimensions of the dimples of fracture surface after each heat treatment considered are greater than the dimples of the corresponding fracture surface just after a stress relieving treatment. The total energy of the fracture is related to the size of the dimples [52]. Brandl et al. have reported that the SLM processing of AlSi10Mg by maintaining the build platform at 300 °C and T6 heat treatment after SLM increase the fatigue resistance. This is due to homogenizing of microstructure after T6 heat treatment [53].

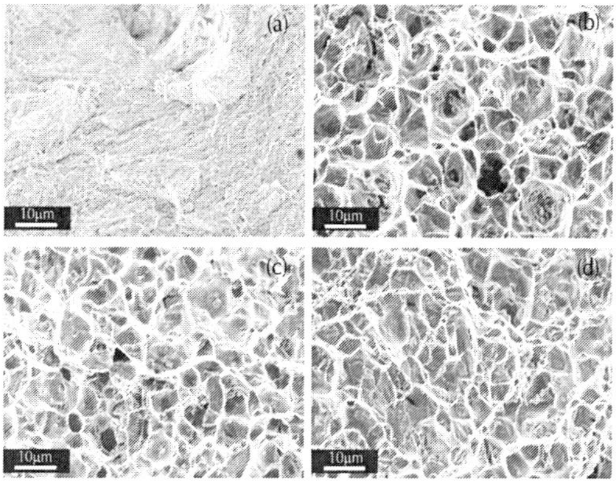

Figure 11: FESEM images at the same magnification of fracture surfaces after DMLS and a stress relieving treatment (a), after an annealing treatment – T2 (b), a T4 treatment (c) and a T6 treatment (d)

DMLS Samples Surface Morphology

Due to its versatility of materials and shapes, the main advantage of DMLS is to produce metal complex-shaped components in one step, but it also has drawbacks that require careful process control: the high temperature gradients and densification ratio during the process yield high internal stresses or part distortion; the risk of balling and dross formation in the melt pool may result in bad surface roughness (from 8 to 20 μm without any post-treatment) [54-55]. The surface finish of a part, defined as the irregularities of any material resulting from machining operations, is critical in many applications, for example those requiring a surface roughness of 0.8 μm or better to avoid premature failure from surface initiated cracking [56]. There is no standard method to enhance the surface quality due to the complex nature of the process and to the different properties of the materials used [54,57]. Laser parameters like laser beam power, spot size and process parameters like scanning speed, hatching distance and layer thickness have a great influence on the quality of the laser sintered samples. The effect of some process parameters on AlSi10Mg parts produced by DMLS was studied through statistically designed experiments based on an *L18* orthogonal array of Taguchi design [12]. Scan speed was found to have the greatest influence on the surface roughness. The Taguchi method uses *S/N* ratio to measure the variations of the experimental design. The equation of *smaller-the-better* was selected for the calculation of *S/N* ratio since it yields the lowest values of surface roughness. It was found that*S/N* ratio is minimized when the scan speed is 900 mm/s, the laser power is 120 W, and the hatching distance is 0.10 mm.

The process of shot-peening involves a mechanical surface treatment whereby small balls impinge on the surface of the component. The repeated impacts not only induce compressive residual stress but also refine the microstructure at the surface and sub-surface region [58,59]. The near surface compressive residual stress field reduces the effective applied stresses of the component during application, which results in delayed crack initiation and retarded early crack propagation. It is important that shot-peening method is optimized to improve the depth of favorable compressive residual stress fields while minimizing surface roughening [59,60].

Figure 12 shows the surface roughness of a sample after the DMLS process (R_a=23.08 µm,R_z=152.92 µm) and then after the shot-peening post processing treatment (R_a=3.35 µm, R_z=31.81 µm). Thanks to the use of a 3D scanner is possible to analyze the entire surface (Figure 12a and 12d) and define the deviation of these from the best-fit 3D plane. The deviation of the sample before and after shot-peening is of 0.17 mm and 0.07 mm respectively (Figure 12b and 12e). Surface topography is classically characterized by surface profiles obtained via electronic contact profilometry (Figure 12 c and 12f).

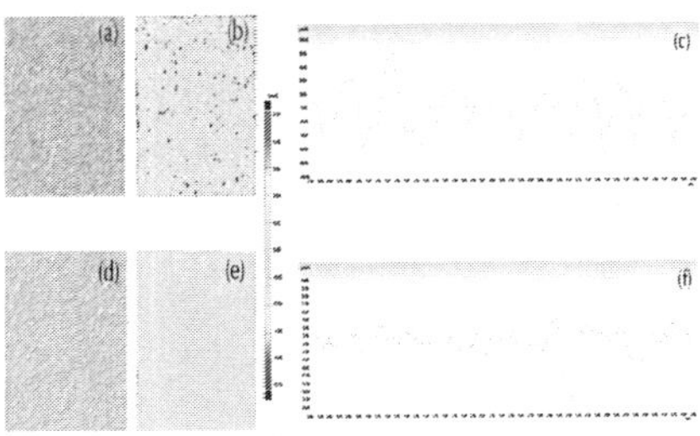

Figure 12: (a) 3D scan surface roughness of a DMLS AlSiMg sample before and (d) after shot-peening. (b) Deviation color maps respect to best fit plane

before and (e) after shot-peening. (c) Roughness profile of the sample before and (f) after shot-peening.

Different roughness parameters are then extracted from the acquired surface profile. Among all the parameters for quantifying surface roughness based on tactile profile sections, R_a, the arithmetical mean deviation of the assessed profile, is by far the most extensive and most used parameter. It possible to see that shot peening with glass beads makes to significantly reduce the surface roughness. Different values of shot peening pressure were analyzed. The best results were obtained using a pressure of 8 bar: the surface roughness was reduced by up to 85%. These considerations on laser sintering parameters and the shot peening effect were confirmed by FESEM observations (Figure 13): the images give evidence of the improvement in surface roughness.

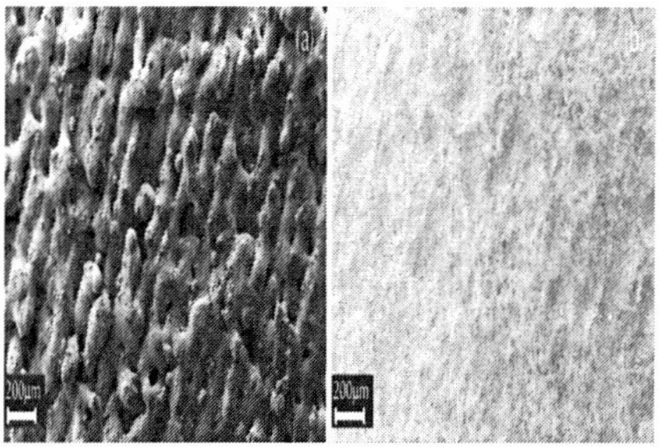

Figure 13: FESEM images of surface roughness of (a) sample with Ra=23 μm and (b) sample after shot peening at 8 bar.

Lightweight Metal Structures by DMLS

One of the main interesting applications of additive manufacturing is for fabrication of customized, lightweight material structures, like periodic cellular lattice structures [61]. The word "cell" derives from the Latin word *cella* meaning a small compartment or enclosed space [62]. When nature constructs things, it is often done from many cells (cellula). Wood and cancellous bone are good examples of this: they are stiff, lightweight and multifunctional. From these natural inspirations a variety of synthetic cellular solids could be made including stochastic foams, consolidated powders, honeycombs and lattice blocks or miniature truss structures. These structures can offer high performance features such as high strength accompanied by a relatively low mass, good energy absorption characteristics and good thermal and acoustic insulation properties, making them suitable for high value aerospace, medical and engineering products [63]. The concept of designed cellular materials is motivated by the desire to put material only where it is needed for a specific application, as nature does. Leonardo da Vinci stated: "In her (nature's) inventions, nothing is lacking, and nothing is superfluous". From a mechanical engineering viewpoint, a key advantage offered by cellular materials is high strength accompanied by a relatively low mass.

In the past 15 years, the area of lattice materials has received considerable research attention due to their inherent advantages over foams in providing light, stiff, and strong materials. Many methods have been developed to analyze various cellular structures. Wang and McDowell have performed a comprehensive review of analytical modeling, mechanics, and characteristics of various metal honeycombs [64,65]. Deshpande et al. [66] have investigated extensively lattice cells, particularly the octet-truss structure.

However, the cellular lattice structures proposed and investigated in the previous studies could not exhibit good manufacturability in SLM. As reported in a recent study, the cellular structures with large unit cell sizes (greater than 5 mm) could not be built using the SLM process because overhanging struts in the cells led to the occurrence of serious deformation [67]. This adds considerable constraints on manufacturing versatile and complex cellular structures to meet requirements of different functions and applications, sacrificing the design freedom of cellular structures and geometrical capability of AM manufacturing. Sacrificial support structures might be used to support overhanging structures and thus prevent deformation. On the counter side, support structures inside complex cellular lattice structures are normally not acceptable because they are very difficult to remove, and waste materials and energy.

It was demonstrated, by the authors chapter, that it is possible to manufacture aluminium lattice structures by DMLS as the ones reported in Figure 14, with desired shape and internal features in a single fabrication step. This was possible because previously many experiments were focused on investigating the limitations of building surfaces without support structures. An overhanging structure is a part of a component that is not supported during building, by solidified material or a substrate on the bottom side. Consequently, the melt pool created by the heat input from the laser is supported by powder material. From this definition, it is clear that a part of a component is an overhanging structure depending on the orientation given to the part while building it.

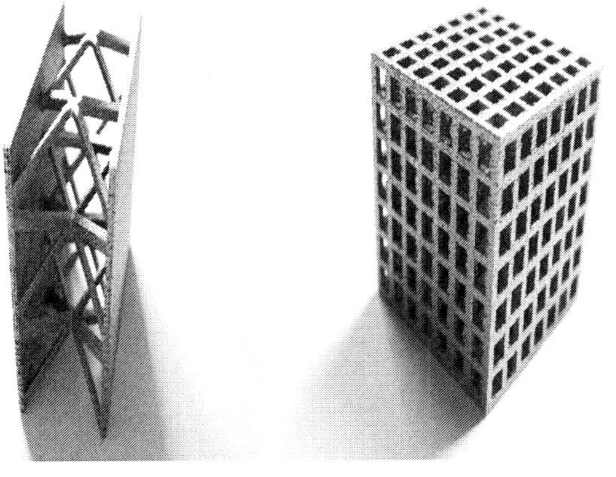

Figure 14: Lightweight components fabricated through DMLS in AlSiMg alloy.

In particular the cubic lattice structure of figure 14 was constructed with an angle of 45 degrees with respect to the building platform.

Following these design rules, a wide variety of different architectures can be made with fine control at the so called "cell level", as shown in Figure 15. The geometrical features selected for investigations were chosen because they could be the building blocks of more complex geometries. The structures have very good base metal properties thanks to the very fine microstructure typical of this process.

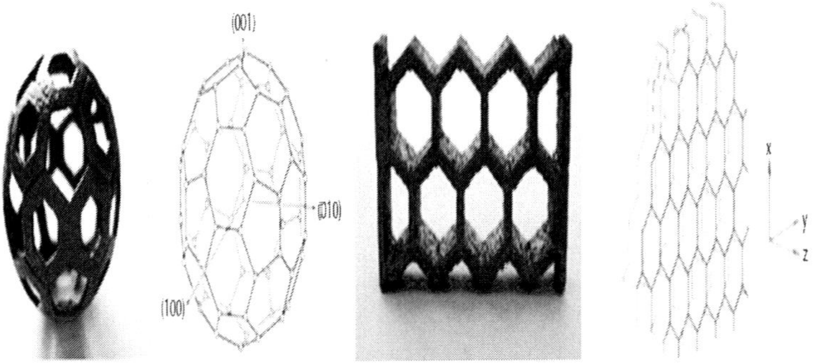

Figure 15: Cellular structures fabricated through DMLS in AlSiMg alloy: a fullerene structure (on the left), and a honeycomb cell structure (on the right).

CASE STUDY — ALUMINIUM MATRIX COMPOSITES (AMCS) BY DMLS

As described in the previous paragraphs, aluminium alloys are currently produced by DMLS, in particular with compositions suitable for casting. The microstructure of these alloys is peculiar, since the very fast cooling occurring after the melting induced by laser provides an ultra-fine microstructure [34] and this effect is responsible for the significant increase of the mechanical properties that is observed with respect to conventional casting processes [37]. In fact, the conventional casting can be followed by a hardening heat treatment to improve its mechanical properties, treatment that seems not necessary in the case of DMLS process. Moreover, as a powder based process, DMLS also provides great opportunity to consolidate second or multiple material particles with metal powders to form novel metal matrix composites (MMCs). There is a growing research to develop MMCs via SLM process [68,69].

In order to be used inside a powder bed system, discontinuous ceramic particles are generally used. Among the most common reinforcement, silicon carbide (SiC) or aluminium oxide (Al_2O_3) are the most used, and in particular SiC-based AMCs are now widely used in many fields: from brake drums and cylinders liners of automobiles, to structural aerospace parts such as rotor vanes and plates. There are several methods to fabricate such composites: powder metallurgy, squeeze casting, stir casting or modified casting; however these conventional techniques are generally not suitable for complex shapes.

In the recent past different research groups investigated the feasibility to employ DMLS to obtain Al-based MMCs. Size and volume fraction of SiC particles have been varied to analyze the behavior of the composite [70], demonstrating that cracking occurs during the preparation by DMLS of these composites. In particular, crack density increases significantly after 15 volume percentage (vol.%) of SiC, and that there was not improvement of wear resistance over 20 vol.% of reinforcement. The same authors, in a previous study, found that microhardness increases with increase of volume fraction of SiC particles, as expected [39]. A couple of years before, Simchi et al. investigated the direct laser sintering of Al–7Si–0.3Mg/SiC composites [71], and showed that the densification rate increases at low SiC fractions but abruptly decreases at >~5 vol.%. Meanwhile, significant reaction occurs between the aluminium melt and the reinforcement particles, leading to formation of $Al_4SiC_4$and silicon particles. However, these studies do not investigate the effects of process parameters on the density of metal matrix composites built by DMLS but the influence of design choices like supports generation and part orientation in the building chamber to avoid the presence of cracks during manufacture.

Starting from AlSiMg alloy, the possibility of composites fabrication by DMLS was investigated. The AlSiMg alloy has a high fluidity in the liquid phase, so that it is possible to prepare almost dense samples. Moreover, if silicon carbide is used a second phase, it is known in literature [72] that a high amount of silicon reduces or suppresses the reaction between aluminium and silicon carbide that brings to the formation of the dangerous aluminium carbide Al_4C_3, generally heavily detrimental for the mechanical properties. For this reason SiC particles were chosen as one of the possible reinforcing materials. -SiC powders from H.C. Starck (UF-15, 15 m^2/g) were used, with density of 3.2 g/ $cm^3$and a mean particle size of 0.55 μm.

Another interesting reinforcement that was tested is the aluminium-magnesium spinel ($MgAl_2O_4$). This oxide has a high melting point of 2315 °C, a high Young's modulus and strength, and low thermal expansion coefficient (CTE). Moreover it is suggested in the literature [73] that it can improve substantially the creep behavior of composites if the size of the particles is very small. Thus nano-sized powders of $MgAl_2O_4$ were used (Nanocerox, mean size around 30 nm)

Since SiC particles are large while $MgAl_2O_4$ particles are much smaller, a different amount of the two reinforcements was used for the two composites, namely 10% and 1% in weight of SiC and $MgAl_2O_4$respectively. To avoid deformation of the alloy spherical particles, simple mixing was done to prepare the composite powders, by using a ball milling system in ceramic jars, without any grinding medium, for 48 hours. The powders were then sieved with a mesh of 63 μm before putting them in the DMLS machine. The parameters used for the preparation of the composites are shown in Table 10.

Table 10: Details of the process parameters employed

Parameters	**AlSiMg**	
	+ SiC	**+ $MgAl_{2O}4$**
Scan speed (mm/s	500,700	600, 800
Laser Power (W)	180,195	195
Hatching dist. (mm)	0.17	0.17, 0.10

In Figure 16 are shown the composite powders observed by FESEM. The metallic powders are not modified during the mixing process, however it is evident that both silicon carbide and nanospinel particles adhere on the surface of the aluminium alloy particles, and no agglomerates of pure ceramic particles are observed. In the case of silicon carbide the particles form a "fluffy" layer (Figure 16a), that causes an increased interaction between the surfaces of the powders. Thus, they have a different behavior in terms of flowability, causing some problems during the spreading of the powders, in particular in the first layers. This phenomenon is evident with silicon carbide, while it does not occur with the nanospinel (Figure 16b), probably due to the very small ceramic grain size and to the lower ceramic content. However it can be seen that the $MgAl_2O_4$ particles cover uniformly the round surface.

In the case of AlSiMg/SiC 10% composites, after a first step of parameter optimization, rectangular samples of 50x10x4 mm dimensions were obtained, with geometrical density between 2.59 and 2.61 g/cm^3. Since the theoretical density of these materials is 2.73 g/cm^3, the corresponding residual porosities are from 4.4 to 5%, a result in line with previous studies [39].

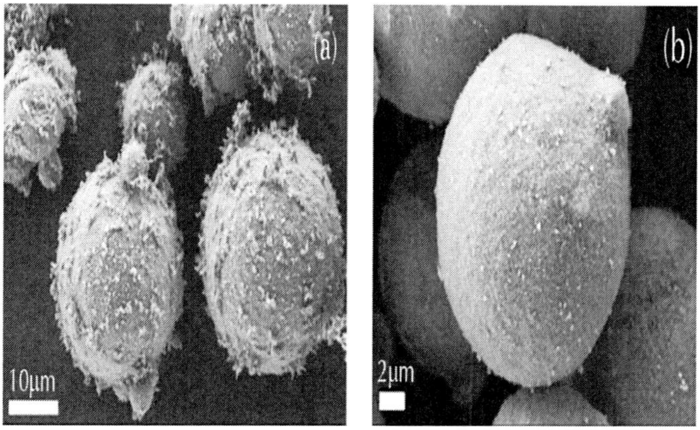

Figure 16: FESEM images of the composites powders: AlSiMg (a) and SiC (10%wt.) and (b) nano-$MgAl_2O_4$(1%wt).

In the case of AlSiMg/SiC 10% composites, after a first step of parameter optimization, rectangular samples of 50x10x4 mm dimensions were obtained, with geometrical density between 2.59 and 2.61 g/cm^3. Since the theoretical density of these materials is 2.73 g/cm^3, the corresponding residual porosities are from 4.4 to 5%, a result in line with previous studies [39]. A problem however arises however with these composites: XRD measurements demonstrate that silicon carbide is almost completely disappeared, while aluminium carbide is formed. This behavior is rather different to what happens for more conventional processes, where 10% Si is sufficient to suppress the Al-SiC reaction [72]. In this case, even if the contact time between molten alloy and ceramic is very low, the temperature is probably very high (in excess of 2000 K) and the reaction occurs. Ghosh et al. [39] already observed this phenomenon, but with a very low silicon content, while in the case of AlSiMg alloy the behavior was somewhat unexpected. The microstructure of the obtained composites is illustrated in the micrograph of Figure 17: it is very fine, as typical of the DMLS process, with elongated acicular grains, probably of Al_4C_3 as suggested by XRD analysis.

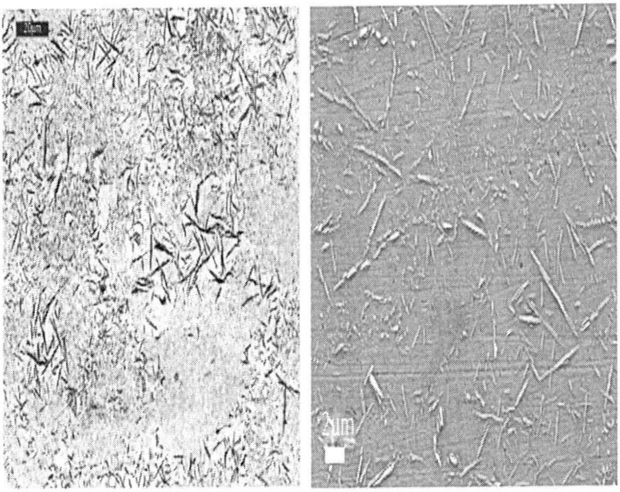

Figure 17: Optical (on the left) and FESEM (on the right) images of AlSiMg/SiC composite microstructure after polishing.

Residual SiC, instead, could be detected only in the very rare cases where some agglomerate is present, suggesting that the well dispersed single silicon carbide particle is instead completely reacted with the matrix. The presence of the ceramic reinforcement on hardness is in any case very evident: Brinell measurements mean value is 178 ± 2 HB for the SiC containing composites, which is 70% higher with respect to pure AlSiMg by DMLS.

In the case of nano-$MgAl_2O_4$ AMCs, the density is more depending on the specific choice of parameters than in the case of silicon carbide containing composites. In this case small samples were obtained to perform a more detailed screening of the parameters, and for this reason samples with 15x15x10 mm size were fabricated. With the parameters optimized in the best range, the measured average densities of the samples varied between 2.59 and 2.63 g/cm^3, corresponding to residual porosities in the range from 2.2 to 3.5% (the theoretical density is 2.689 g/cm[3]).

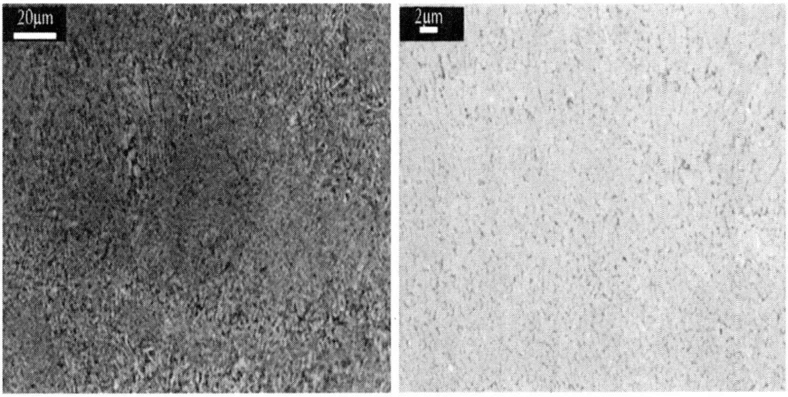

Figure 18: Optical (on the left) and FESEM (on the right) images of AlSiMg/Mg-nanospinel composite microstructures.

In this case, the XRD analysis shows the presence of Al, Si, with some extremely small peaks that were attributed to the inter-metallic phase Mg_2Si and to some mixed metal-oxide of unknown origin ($AlSiO_x$), with a spectrum that is rather similar to the one of the pure alloy. Indeed, in this case the XRD technique is clearly less powerful to recognize possible reactions, due to the low content of reinforcing ceramic. The microstructure of these composites is also very similar to the case of the pure aluminium matrix, as illustrated in the micrographs of Figure 18. As typical of the DMLS process the microstructure is very fine, made by submicrometric grains, elongated in the correspondence of the melt pools contour due to the heat transfer flux direction. A careful observation however shows a higher inhomogeneity of these composites with respect to the pure aluminium alloy, suggesting that the presence of nano-$MgAl_2O_4$ particles changes the behavior during solidification. This inhomogeneity, together with the higher residual porosity, reflects on the values of Brinell Hardness, that is 93 ± 3 HB, 11% lower than the pure AlSiMg alloy processed through DMLS.

ACKNOWLEDGEMENTS

The authors wish to acknowledge financial support from the AMAZE Project, which is co-funded by the European Commission in the 7th Framework Programme (contract FP7-2012-NMP-ICT-FoF-313781), by the European Space Agency and by the individual partner organizations.

REFERENCES

1. Gibson I, Rosen WD, Stucker B. Additive Manufacturing Technologies-Rapid Prototyping to Direct Digital Manufacturing. Springer; 2010.
2. Wohlers T. Wohlers Report 2011 State of the Industry, Annual Worldwide Progress Report. Wohlers Associates; 2011.
3. Campbell I, Bourell D, Gibson I. Additive manufacturing: Rapid prototyping comes of age. Rapid Prototyping Journal 2012; 18 (4): 255–258.
4. http://www.amaze-project.eu/
5. http://www.rapolac.eu/
6. Gerrits A, Jones CL & Valero R. Custom-Fit: Quality of Life of European Sporting Public through Custom-Fit Products. The Engineering of Sport. 2006, 6: 5-10.
7. http://www.e-break.eu/
8. http://www.tialcharger.de/
9. Wohlers T. Additive Manufacturing Advances. Manuf. Eng. 2012; 148 (4): 55–63
10. Hot-Fire Tests Show 3-D Printed Rocket Parts Rival Traditionally Manufactured Parts. 2013. [Online]. Available: http://www.nasa.gov/exploration/systems/sls/3dprinting.html [Accessed: 13-Aug-2013].
11. ASTM F2792-12a: Standard Terminology for Additive Manufacturing Technologies. DOI:10.1520/F2792-12A
12. Calignano F, Manfredi D, Ambrosio EP, Iuliano L, Fino P. Influence of process parameters on surface roughness of aluminium parts produced by DMLS. Int. J. Adv. Manuf. Technol. 2013; 67 (9–12): 2743–2751.

13. Aumund-Kopp C, Petzoldt F. Laser Sintering of parts with complex internal structures. In: Proceedings of the 2008 world congress on powder metallurgy & particulate materials. 2008; 1 (3): 385–397.
14. Tsopanos S, Mines RAV, McKown S, Shen Y, Cantwell WJ, Brooks W, Sutcliffe CJ. The Influence of Processing Parameters on the Mechanical Properties of Selectively Laser Melted Stainless Steel Microlattice Structures. J. Manuf. Sci. Eng. 2010; 132 (4): 041011.
15. Childs THC, Hauser C, Badrossamay M. Selective laser sintering (melting) of stainless and tool steel powders: experiments and modelling. Proc. Inst. Mech. Eng. Part B J. Eng. Manuf. 2005; 219 (4): 339–357.
16. Tolosa I, Garciandía F, Zubiri F, Zapirain F, Esnaola A. Study of mechanical properties of AISI 316 stainless steel processed by 'selective laser melting', following different manufacturing strategies. Int. J. Adv. Manuf. Technol. 2010; 51 (5–8): 639–647.
17. Li R, Liu J, Shi Y, Du M, Xie Z. 316L stainless steel with gradient porosity fabricated by selective laser melting. J. Mater. Eng. Perform. 2010; 19 (5): 666–671.
18. Averyanova M, Cicala E, Bertrand P, Grevey D. Experimental design approach to optimize selective laser melting of martensitic 17-4 PH powder: part I – single laser tracks and first layer. Rapid Prototyping Journal. 2012; 18 (1): 28–37.
19. Dewidar MM, Dalgarno KW, Wright CS. Processing conditions and mechanical properties of high-speed steel parts fabricated using direct selective laser sintering. Proc. Inst. Mech. Eng. Part B J. Eng. Manuf. 2003; 217 (12): 1651–1663.
20. Yasa E, Deckers J, Kruth JP, Rombouts M, Luyten J. Charpy impact testing of metallic selective laser melting parts. Virtual Phys. Prototyp. 2010; 5 (2): 89–98.
21. Li X, Wang C, Zhang W, Li Y. Fabrication and compressive properties of Ti6Al4V implant with honeycomb-like structure for biomedical applications. Rapid Prototyping Journal. 2010; 16 (1): 44–49.
22. Vrancken B, Thijs L, Kruth JP, Van Humbeeck J. Heat treatment of Ti6Al4V produced by Selective Laser Melting: Microstructure and mechanical properties. J. Alloys Compd. 2012; 541: 177–185.

23. Van Hooreweder B, Moens D, Boonen R, Kruth JP, Sas P. Analysis of Fracture Toughness and Crack Propagation of Ti6Al4V Produced by Selective Laser Melting. Adv. Eng. Mater. 2012; 14 (1–2): 92–97.

24. Song B, Dong S, Zhang B, Liao H, Coddet C. Effects of processing parameters on microstructure and mechanical property of selective laser melted Ti6Al4V. Mater. Des. 2012; 35: 120–125.

25. Marcu T, Todea M, Gligor I, Berce P, Popa C. Effect of surface conditioning on the flowability of Ti6Al7Nb powder for selective laser melting applications. Appl. Surf. Sci. 2012; 258 (7): 3276–3282.

26. Wang Z, Guan K, Gao M, Li X, Chen X, Zeng X. The microstructure and mechanical properties of deposited-IN718 by selective laser melting. J. Alloys Compd. 2012; 513: 518–523.

27. Yadroitsev I, Pavlov M, Bertrand P, Smurov I. Mechanical properties of samples fabricated by selective laser melting. In: 14èmes Assises Européennes du Prototypage & Fabrication Rapide, 2009.

28. Kumar S, Pityana S. Laser-based additive manufacturing of metals. Adv. Mater. Res. 2011; 227: 92–95.

29. Kumar S. Selective Laser Sintering: Recent Advances. In: 4th Pacific International Conference on Applications of Lasers and Optics. Wuhan-China, 2010.

30. Khan M, Dickens PM. Processing parameters for Selective Laser Melting (SLM) of gold. In: Proceedings of Solid Freeform Fabrication symposium, 2008: 278–289.

31. Louvis E, Fox P, Sutcliffe CJ. Selective laser melting of aluminium components. J. Mater. Process. Technol. 2011; 211 (2): 275–284.

32. Olakanmi EO, Dalgarno KW, Cochrane RF. Laser sintering of blended Al-Si powders. Rapid Prototyping Journal. 2012; 18 (2): 109–119.

33. Olakanmi EO, Cochrane RF, Dalgarno KW. Densification mechanism and microstructural evolution in selective laser sintering of Al – 12Si powders. J. Mater. Process. Technol. 2011; 211 (1): 113–121.

34. Thijs L, Kempen K, Kruth JP, Van Humbeeck J. Fine-structured aluminium products with controllable texture by selective laser melting of pre-alloyed AlSi10Mg powder. Acta Mater. 2013; 61 (5): 1809–1819.

35. Kempen K, Thijs L, Yasa E, Badrosamay M, Verheecke W, Kruth JP. Process optimization and microstructural analysis for selective laser melting of AlSi10Mg. In: Proceedings of Solid Freeform Fabrication symposium, 2011: 484–495.

36. Manfredi D, Calignano F, Ambrosio EP, Krishnan M, Canali R, Biamino S, Pavese M, Atzeni E, Iuliano L, Fino P and Badini C. Direct Metal Laser Sintering: an additive manufacturing technology ready to produce lightweight structural parts for robotic applications. La Metallurgia Italiana. 2013; 105 (10): 15-24.

37. Manfredi D, Calignano F, Krishnan M, Canali R, Ambrosio EP, Atzeni E. From Powders to Dense Metal Parts: Characterization of a Commercial AlSiMg Alloy Processed through Direct Metal Laser Sintering. Materials. 2013; 6 (3): 856-869.

38. Kumar S, Kruth JP. Composites by rapid prototyping technology. Mater. Des. 2010; 31 (2): 850–856.

39. Ghosh SK, Saha P, Kishore S. Influence of size and volume fraction of SiC particulates on properties of ex situ reinforced Al-4.5Cu-3Mg metal matrix composite prepared by direct metal laser sintering process. Mater. Sci. Eng. A, 2010; 527 (18–19): 4694–4701.

40. Thijs L, Verhaeghe F, Craeghs T, Humbeeck JV, Kruth JP. A study of the microstructural evolution during selective laser melting of Ti-6Al-4V. Acta Mater. 2010; 58 (9): 3303–3312.

41. Spierings AB, Herres N, Levy G. Influence of the particle size distribution on surface quality and mechanical properties in AM steel parts. Rapid Prototyping Journal. 2011; 17 (3): 195–202.

42. Buchbinder D, Schleifenbaum H, Heidrich S, Meiners W, Bültmann J. High Power Selective Laser Melting (HP SLM) of Aluminum Parts. Phys. Procedia 2011; 12: 271–278.

43. Qiu C, Adkins NJE, Attallah MM. Microstructure and tensile properties of selectively laser-melted and of HIPed laser-melted Ti-6Al-4V. Mater. Sci. Eng. A, 2013; 578: 230–239.

44. Thijs L, Montero Sistiaga ML, Wauthle R, Xie Q, Kruth JP, Van Humbeeck J. Strong morphological and crystallographic texture and resulting yield strength anisotropy in selective laser melted tantalum. Acta Mater., 2013; 61 (12): 4657–4668.

45. Kruth JP, Badrosamay M, Yasa E, Deckers J, Thijs L, Van Humbeeck J. Part and material properties in selective laser melting of metals. In: Proceedings of the 16th International Symposium on Electromachining (ISEM XVI), 2010.

46. Shiomil M, Yamashital T, Osakada K, Shiomi M, Yamashita T, Abe F, Nakamura K. Residual Stress within Metallic Model Made by Selective Laser Melting Process. Ann. CIRP, 2014; 53 (1): 195–198.

47. Zaeh MF, Branner G. Investigations on residual stresses and deformations in selective laser melting. Prod. Eng., 2009; 4 (1): 35–45.

48. Kruth JP, Yasa E, Deckers J. Roughness improvement in Selective Laser Melting. In: Proceedings of the 3rd International Conference on Polymers and Moulds Innovations. PMI, 2008: 170–183.

49. ASM Handbook, Volume 2-Properties and Selection: Non ferrous Alloys and Special-Purpose Materials, and Volume 4 – Heat treating. 1990, ASM International The Materials Information Company, United States of America, ISBN 0-87170-379-3.

50. Hauser C, Childs THC, Badrossamay M. Further developments in process mapping and modelling in direct metal selective laser melting. In: Proceedings of the Solid Freeform Fabrication symposium, 2004; 448–459

51. Krishnan M. Investigation of material and mechanical properties of Al alloy and Al based MMC parts produced by DMLS for industrial application. PhD Thesis. Politecnico di Torino; 2014.

52. Hertzberg RW. Deformation and Fracture Mechanics of Engineering Materials. John Wiley & Sons Inc, 1996.

53. Brandl E, Heckenberger U, Holzinger V, Buchbinder D. Additive manufactured AlSi10Mg samples using Selective Laser Melting (SLM) Microstructure, high cycle fatigue, and fracture behavior. Materials and Design. 2012; 34: 159-169.

54. Delgado J, Ciurana J, Rodríguez CA. Influence of process parameters on part quality and mechanical properties for DMLS

and SLM with iron-based materials. Int J Adv Manuf Technol. 2012; 60:601–610

55. Kruth JP, Levy G, Klocke F, Childs THC. Consolidation phenomena in laser and powder-bed based layered manufacturing. CIRP 2007; 56(2):730–759

56. Dalgarno K. Materials research to support high performance RM parts. Rapid Manufacturing 2nd International Conference, Loughborough University, 2007; 147–56

57. Sachdeva A, Singh S, Sharma VS. Investigating surface roughness of parts produced by SLS process. Int J Adv Manuf Technol. 2012.

58. Evans A, Bruno G. Relaxation of residual stress in shot peened Udimet 720Li under high temperature isothermal fatigue. Int J Fatigue. 2005; 27: 1530–1534.

59. Turski M, Clitheroe S, Evans AD, Rodopoulos C. Engineering the residual stress state and microstructure of stainless steel with mechanical surface treatments. Appl Phys A. 2010; 99: 549–56.

60. Murr LE, Gaytan SM, Ramirez DA, Martinez E, Hernandez J, Amato KN, Shindo PW, Medina FR, Wicker RB. Metal Fabrication by Additive Manufacturing Using Laser and Electron Beam Melting Technologies. J. Mater. Science Technology. 2012; 28(1): 1–14.

61. Ashby MF et al., Metals Foams: A Design Guide. Woburn, MA: Butterworth-Heinemann, 2000.

62. Gibson LJ, Ashby MF. Cellular Solids, Structure and Properties. 2nd Edition, Cambridge University Press, Cambridge, 1997.

63. Williams CB, Cochran JK, Rosen DW. Additive manufacturing of metallic cellular materials via three-dimensional printing. Int. J. Adv. Manuf. Technol. 2011; 53: 231–239.

64. Wang AJ, McDowell DL, Optimization of a metal honeycomb sandwich beam-bar subjected to torsion and bending. Int. J. Solids Struct. 2002; 40(9): 2085-2099.

65. Wang AJ, McDowell DL. Yield surfaces of various periodic metal honeycombs at intermediate relative density. Int. J. Plasticity. 2005; 21(2): 285-320.

66. Deshpande VS, Fleck NA, Ashby MF. Effective properties of the octet-truss lattice material. Journal of Mechanics and Physics of Solids. 2001; 49(8): 1747-1769.

67. Yan C, Hao L, Hussein A, Raymont D. Evaluations of cellular lattice structures manufactured using selective laser melting. International Journal of Machine Tools & Manufacture. 2012; 62: 32–38.

68. Kumar S, Kruth JP. Composites by rapid prototyping technology. Materials and Design. 2010; 31: 850-856.

69. Hao L, Dadbakhsh S. Materials and process aspects of selective laser melting of metals and metal matrix composites: a review. Chinese Journal of Lasers. 2009; 36 (12): 3192–3203.

70. Ghosh SK, Saha P. Crack and wear behavior of SiC particulate reinforced aluminium based metal matrix composite fabricated by direct metal laser sintering process. Materials and Design. 2011; 32: 139-145.

71. Simchi A, Godlinski D. Effect of SiC particles on the laser sintering of Al–7Si–0.3Mg alloy. Scripta Materialia. 2008; 59: 199-202.

72. Narciso J, García-Cordovilla C, Louis E. Reactivity of thermally oxidized and unoxidized SiC particulates with aluminium-silicon alloys. Materials Science and Engineering B. 1992; 15: 148-155

73. De Vita A, Roitti S, Sbaizero O. Creep of aluminium metal matrix composites reinforced with magnesium spinel. Key Engineering Materials. 2004; 264-268: 953-956.

Chapter 8

Zinc Oxide — Linen Fibrous Composites: Morphological, Structural, Chemical, Humidity Adsorptive and Thermal Barrier Attributes

Narcisa Vrinceanu[1,2], Alina Brindusa Petre[1], Claudia Mihaela Hristodor[1], Eveline Popovici[3], Aurel Pui[1], Diana Coman[2] and Diana Tanasa[1]

[1] "Al.I.Cuza" University of Iasi, Iasi, Romania

[2] "L.Blaga" University of Sibiu, Romania

[3] Al.I.Cuza" University of Iasi, Faculty of Chemistry, Department of Materials Chemistry, Romania

INTRODUCTION

The augmented requirement for fibrous supports (yarns) possessing multifunctionality implies powerful emerging multidisciplinary approaches as well as the connection with the traditional scientific disciplines [1]. Finishing processes through nanoparticles were among the first commercial application in textiles domain.

Due to poor fixing of these nanoparticles on the textile surface, these finishes were not resistant to washing. Nanofinishings with improved bonding properties in fabrics and also impart desired wettability will result by using hydrophobic/hydrophilic functional polymer fibrous matrices as dispersion medium for nanoparticles.

Nanoparticles are extremely reactive, due to their high surface energy, and most systems undergo aggregation without protection of their surfaces. To eliminate or minimize generated waste and implement sustainable processes, recently green chemistry and chemical processes have been emphasized for the preparation of nanoparticles [2]. Much attention is now being focused on polysaccharides used as the protecting agents of nanoparticles. As stabilizing agent soluble starch has been selected and as the reducing agent in aqueous solution of $AgNO_3$ for silver nanoparticle growth, and a-D-glucose, has been elected. To maintain noble metal (platinum, palladium and silver) nanoparticles in colloid suspension, Arabinogalactan has been used as a novel protecting agent for [3]. Synthesized platinum, palladium and silver nanoparticles with narrow size distribution have been achieved by using porous cellulose fibers as the stabilizer [4]. Pt nanoparticles can catalyze the carbonization of cellulose and mesoporous amorphous carbon is fabricated in high yields. The results are carbon-based functional composites with metal nanoparticles, showing that self-supporting macroporous sponges of silver, gold and copper oxide, as well as composites of silver/copper oxide or silver/titania can be routinely prepared by heating metal–salt-containing pastes of dextran, chosen as a soft template [5, 6]. Polysaccharides could be used as stabilizer to synthesize nanoparticles of metal oxide and sulfides. Zinc oxide nanoparticles can be synthesized using water as a solvent and soluble starch as a stabilizer [7-9] while CdS nanoparticles have been prepared in a sago starch matrix.

In an earlier study, ZnO nanoparticles synthesis can be made with the assistance of MCT- -CD (monochlorotriazinyl– -cyclodextrin) by using a sol-gel method [10]. MCT- -CD, a commercially available -cyclodextrin with a reactive monochlorotriazinyl group, is used as a stabilizer. The so called anchor group reacting with cellulose hydroxyl radicals and cyclodextrin molecule is covalently bonded, to the fiber surface. The stable bound of cyclodextrin onto the textile fibers allows its properties to become intrinsic to the modified supports, thus a new generation of *intelligent textiles possessing* enhanced sorption abilities/capacities, as well as possessing active molecules release wasborn. Besides, as polysaccharide, MCT- -CD shows interesting dynamic supramolecular associations facilitated both by inter- and intra-molecular hydrogen bonding, and polar groups. When a material is exposed to environmental water vapors, the water molecules firstly reacts with surface polar groups, forming a molecular monolayer.

Zinc oxide (ZnO), an n-type semiconductor, is a very interesting multifunctional material and has promising applications in solar cells, sensors, displays, gas sensors, varistors, piezoelectric devices, electro-acoustic transducers, photodiodes and UV light emitting devices. The adhesion between the ZnO nanoparticles and polymer through simple wet chemical method is rather poor and the nanoparticles may be removed from the host easily. In light of this, it is believed that the hydrothermal method can be a more promising way for fabricating nanomaterials because it can be used to obtain products with modified morphological and chemical attributes with high purity. as well as stability in terms of water vapour sorption-desorption. Zn^{2+} ions can penetrate into the interior of linen fibrous support (fabric) easily when soluble salt such as zinc acetate ($Zn(OAC)_2$) is used. Reaction of Zn^{2+}ions leads to crystallization of ZnO nanoparticles within the linen fabric and to the formation of an encapsulated complex in the hydrothermal environment. The formation procedure can be described through two steps as shown in Fig. 1. Firstly, coordination compounds are formed through chelation between Zn^{2+} ions and the hydroxyl groups of linen fabric. Secondly, the in-situ crystallization of Zn chelate complex occurs under the hydrothermal treatment and forms a ZnO-coated linen fabric. The ZnO nanoparticles can thus be attached firmly within the linen fiber surface.

ADVANCES IN ZNO SYNTHESIS

The idea of the interaction of materials with water vapors is a new area of research. Almost all materials have some interaction with moisture that is present in their surroundings. The effects of water can be both harmful and beneficial depending on the material and how it is used. Consequently, the point to determine a correlation between morphological, structural and chemical characterization and the water vapor sorption behavior of the analyzed samples has been emphasized. Consequently, the obtained textiles should find their applicability in textile processing industry subdomains, where a certain level of hydrophilicity/hydrophobicity is mandatory.

The main cause of polymeric materials degradation is the exposure to various factors such as: heat, UV light, irradiation ozone, mechanical stress and microbes. Degradation is promoted by oxygen, humidity and strain, and results in such flaws as brittleness, cracking, and fading [11-13]. There have been research reports targeting nanosized magnetic materials synthesis, having significant potential for many applications.

The applications of ZnO particles are numerous: varistors and other functional devices, reinforcement phase, wear resistant and anti-sliding phase in composites due to their high elastic modulus and strength. Otherwise, ZnO particles exist in anti-electrostatic or conductive phase due to their current characteristics. Few studies have been concerned with the application of ZnO nanoparticles in coatings system with multi-properties. The nano-coatings can be obtained by the traditional coatings technology, i.e., by filling with nanometer-scale materials by filling with nano-materials, both structure and functional properties of coatings can be modified. Super-hardness, wear resistant, heat resistance, corrosion resistance, and about function, anti-electrostatic, antibacterial, anti-UV and infrared radiation all or several of them can be realized.

Another idea this paper review was centered to was to study the thermal degradation behavior of some textile nanocomposites made of nano/micron particle grade zinc oxide and linen fibrous supports, and to discuss the thermal degradation mechanism of the above mentioned structures. There is also potential to highlight the effect of the functionalization agent - MCT- -CD (monochlorotriazinyl– – cyclodextrin) on the thermal stability and degradation mechanism of ZnO nanocoated linen fibrous samples.

In order to characterize the surface morphology and chemical composition of the treated supports, instrumental methods were conducted to measure the particle sizes of the reduced zinc oxide particles. The understanding of the thermal behavior of these fibers is very important since in general several conventional techniques used in textile processing industry, are conducted at high temperature.

The MCT- -CD (monochlorotriazinyl– eta-cyclodextrin) under the trade name CAVATEX or CAVASOL® W7 MCT (CAVATEX) from Wacker Chemie AG, Zn $(OAc)_2$, with an assay of 97%, urea and acetic acid (assay 99%) from CHIMOPAR, cetyltrimethylammonium bromide (CTAB) from Merck Company, with an assay of 97% were utilized

Two 100 % twill linen desized, scoured and bleached supports, each of size 3 cm × 3 cm were used as fibrous support. One of the supports has been coated with a certain concentration of MCT- -CD (monochlorotriazinyl– -cyclodextrin) [14-17].

Table 1: Synthesis conditions for each of the sample

Sample	**Specifications**
Reference 1	linen fibrous support
Reference 2	ZnO powder hydrothermally synthesized, non-calcinated
Sample 1	Functionalization of linen support with MCT- –CD (**M**ono**C**hloro**T**riazinyl– -**C**yclo**D**extrin) by exhaustion and thermal treatment
Sample 3	ZnO powder hydrothermally synthesized onto linen fibrous support
Sample 4	ZnO powder hydrothermally synthesized onto functionalized linen fibrous support
Sample 5	ZnO powder hydrothermally synthesized onto functionalized linen fibrous with the assistance of CTAB (Cetyl TrimethylAmmonium Bromide)
Sample 6	ZnO powder hydrothermally synthesized onto functionalized linen fibrous with the assistance of P123 (**M**ono**C**hloro**T**riazinyl- – **C**ycloDextrin)
Sample 7	MCT- –CD (**M**ono**C**hloro**T**riazinyl– -**C**ycloDextrin)

Fundamental Technique for Synthesizing and Characterizing Nano-ZnO_Particles

The review was focused onto the fibrous supports (yarns) previously grafted/functionalized with MCT- -CD. The grafting process of the textile fabric was performed following two other processes: the exhaustion and squeezing treatment and the heat treatment at 160°C. The purpose of these two treatments was the grafting the linen [18-20].

ZnO nanoparticles were synthesized *in-situ* on linen fibrous supports (yarns) having a certain concentration of MCT- -CD by using the hydrothermal method. The linen samples with sizes of 30 x30 cm^2 were immersed in the solution prepared as follows: zinc acetate $Zn(CH_3COO)_2 . 2H_2O$, purity – 99%) (0,005 mol/1000mL) as precursor was solved in de-ionized water to form a uniform solution by stirring and then 0, 1 mol of urea solution was added drop-wise with constant stirring. Second, the pH value of the mixed solution was adjusted to 5 by adding acetic acid drop wise. The final reaction mixture was then vigorously stirred for two hours at room temperature and poured into 100 mL stainless-steel autoclaves made of Teflon (poly[tetrafluoroethylene), followed by immersion of the fibrous supports (yarns). Then the autoclaves were placed in the oven for the hydrothermal treatment at 90°C overnight. The autoclaves were then cooled down to room temperature. The treated fabrics were then removed from the autoclaves. The treated fabrics were washed several times with distilled water. After complete washing the composites were dried at 60◦C overnight for complete conversion of the remaining zinc hydroxide to zinc oxide.

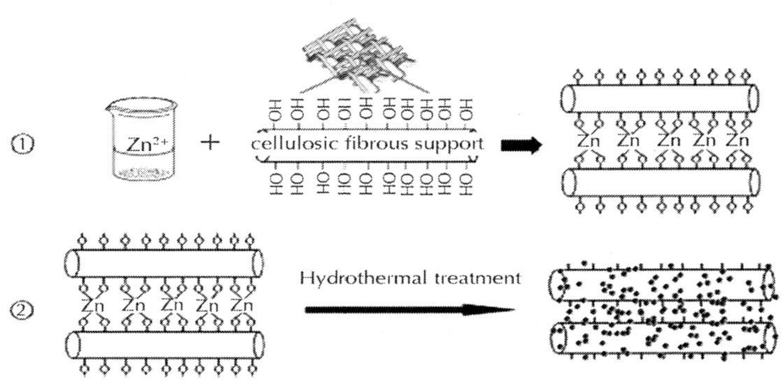

Figure 1: Flow chart for the preparation of nanoparticle coated linen support [10].

Thermal treatment relied into two main stages, into the calcination oven. Firstly, the samples were subjected to an increasing of temperature up to 150°C; secondly the probes were heated up to 350°, 450° respectively.

Scanning Electron Microscope (SEM) images were acquired with a Quanta 200 3D Dual Beam type microscope, from FEI Holland, coupled with an EDS analysis system manufactured by EDAX - AMETEK Holland equipped with a SDD type detector (silicon drift detector). Taking into account the sample type, the analyses have been performed, using Low Vacuum working mode, (as in High Vacuum working type). Both for the acquisition of secondary electrons images (SE – secondary electrons) and EDS type elemental chemical analyses, LFD (Large Field Detector) type detector was used, running at a pressure of 60 Pa, and a voltage of 30 kV.

The ZnO–MCT- -CD treated fabrics were tightly packed into the sample holder. X-ray Diffraction (XRD) data for structural characterization of the various prepared samples of ZnO were collected on an X-ray diffractometer (PW1710) using Cu-K radiation (k = 1.54 Å) source (applied voltage 40 kV, current 40 mA). About 0.5 g of the dried particles were deposited as a randomly oriented powder onto a Plexiglass sample container, and the XRD patterns were recorded at 2 angles between 20 ° and 80°, with a scan rate of 1.5°/min. Radiation was detected with a proportional detector [21-25].

Evaluation of Crystallinity

The extent of crystallinity (*Ic*) was estimated by means of Eq. (1), where I_{020} is the intensity of the 020 diffraction peak at 2 angle close to 22.6°, representing the crystalline region of the material, and I_{am} is the minimum between 200 and 110 peaks at 2 angle close to 18°, representing the amorphous region of the material in cellulose fibres [26-28]. I_{020} represents both crystalline and amorphous materials while I_{am} represents the amorphous material.

IC = I020−IamI020 x 100(%) (1)

A *shape factor* is used in x-ray diffraction to correlate the size of sub-micrometre particles, or crystallites, in a solid to the broadening of a peak in a diffraction pattern. In the Scherrer equation,

$$\tau = K \cdot \lambda \beta \cos\theta$$

where *K* is the shape factor, is the x-ray wavelength, is the line broadening at half the maximum intensity (FWHM) in radians, and is the Bragg angle [29]. is the mean size of the ordered (crystalline) domains, which may be smaller or equal to the grain size. The dimensionless shape factor has a typical value of about 0.9, but varies with the actual shape of the crystallite.

FTIR was used to examine changes in the molecular structures of the samples. Analysis has been recorded on a FTIR JASCO 660+ spectrometer. The analysis of studied samples was performed at 2 cm^{-1} resolution in transmission mode. Typically, 64 scans were signal averaged to reduce spectral noise.

For the studied samples dynamic vapours sorption (DVS) capacity, at 25°C averaging in the domain of relative humidity (RH) 0-90% has been investigated by using an IGAsorp apparatus, a fully automated gravimetric analyzer, supplied by Hidden Analytical, Warrington - UK). It is a standard sorption equipment, which has a sensitive microbalance (resolution 1µg and capacity 200 mg), which continuously registers the weight of the sample in terms of relative humidity change, at a temperature kept constant by means of a thermostatically controlled water bath. The measuring system is controlled by appropriate software.

To study water sorption at atmospheric pressure, a humidified stream of gas is passed over the sample.

The differential scanning calorimetry (DSC) analysis of fibrous supports - ZnO composites were carried out using a NETZSCH DSC 200 F3 MAIA instrument under nitrogen. Initial sample weight was set as 30-50 mg for each operation. The specimen was heated from room temperature to 350°C at a heating rate of 10°C/min.

PROMINENT ASSESSED FEATURES OF FIBROUS COMPOSITES

From the obtained images it was clearly distinguished the hexagonal shape of ZnO agglomerations and the morphology of linen fibers (Fig.2c).

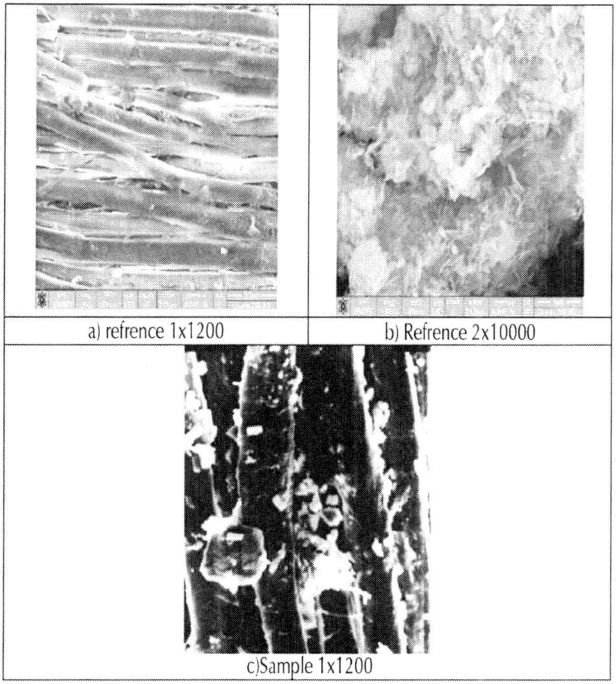

a) refrence 1x1200 b) Refrence 2x10000 c)Sample 1x1200

Figure 2: SEM images of: reference samples and of the functionalized linen support with MCT- –CD sample [10].

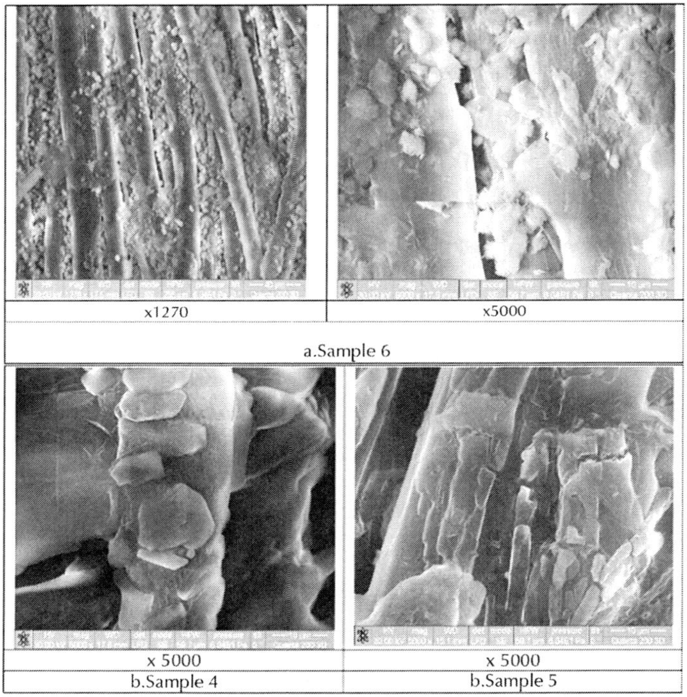

Figure 3: Images of some textile composites [10].

The SEM images of functionalized linen supports coated with ZnO with assistance of the studied surfactants (Fig 3 a and b) indicate different shapes of deposited ZnO.

On the other hand, ZnO nanoparticles exhibited hexagonal form like flowers of ZnO nanocrystals, if the treatment was assisted by P123 surfactant (Fig. 3 a) and lamellar morphology if the treatment was assisted by CTAB (Fig. 3 b) respectively.

The particles uniformly cover the fibrous support surface and as results, the fibrous supports surface became coarser after the treatment.

The adhesion strength of ZnO particles on fibrous support is different in terms of the applied surfactant treatment and was tested after repeated washing cycles (1 minute ten times).

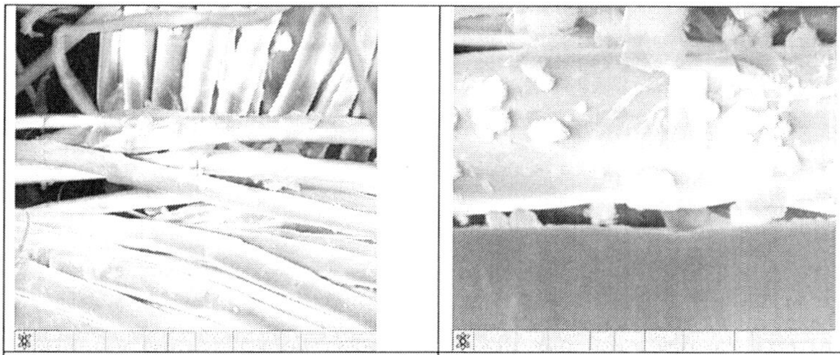

(a) x1200)	(b) (x5000)
sample 3	sample 6

Figure 4: SEM images of some textile composites after repeated washing cycles [10].

According to the SEM images (Fig 4 a and b), the adhesion strength of ZnO powder hydrothermally deposited onto functionalized linen fibrous support is superior in the case of functionalized surface (Fig 4 b) compared with the non-functionalized surface (Fig 4a). The functionalization advantage has been evaluated considering the durability of ZnO on the support surface after repeated cycles of washing. After washing the coating particles fell off easily for the ZnO powder hydrothermally synthesized without functionalization, which might have been caused by the weak attaching force.

As shown in Fig. 4, before treatment the diameters of fibrous supports (individual yarns) are about 10 - 20 μm; after treatment SEM image show very clearly the individual yarns, covered by various ZnO aggregates deposition.

Mechanistic Aspect of Nanoparticle Formation

The shape and the manner of covering depend of linen grafting agent assistance. This result is correlated with the high number of coordinating functional groups (hydroxyl and glucoside groups) of the MCT- -CD which can form complexes with divalent metal ions [15]. During the synthesis time, it might be possible that the majority of the zinc ions were closely associated with the MCT- -CD molecules. Based on the previous research, it can be claimed that nucleation and initial crystal growth of ZnO may preferentially occur on MCT- -CD [16]. Moreover, as polysaccharide, MCT- -CD showed interesting dynamic supramolecular associations facilitated by inter- and intra-molecular hydrogen bonding, which could act as matrices for nanoparticle growth in size of about 30–40 nm. They aggregated to irregular ZnO–CMC nanoparticles in a further step (Figures 4a) and 4 b). In these figures, SEM images of linen supports coated with ZnO with assistance of the two surfactants show that the nanoparticles exhibited an approximately lamellar morphology and the particles can be seen to be coated on the fibrous support surface (yarn). As result, the fibrous supports (yarns) surface became coarser after the treatment.

In the case of CTAB assistance, on the yarns surfaces large ZnO particles, covering the yarn as a bark are noticeable (Fig. 3b), involving that the large particles may be formed via precipitation followed by a step-like aggregation process. In addition, according to the SEM images of the coated fabric, the uniformity of the fabric coated with ZnO powder hydrothermally synthesized with assistance of CTAB (Cetyl trimethylammonium bromide) is better than that of ZnO powder hydrothermally synthesized in the presence of Pluronic P123 and possesses good washing fastness. The last one has not been measured, but it has *apriori* been evaluated. This phenomenon can be explained by the fact that the repeated cycles of washing and rinsing did not conduct to the washing away of the ZnO particles; subsequently the zinc oxide proven a low extent of washing fastness. This statement is also in a good correlation with the XRD results, claiming a slight shift of ZnO intensity peaks, meaning that the nucleation of the zinc oxide occurred not only the support surface, but also within the nanocavities, due the fibers roughness.

In case of ZnO powder hydrothermally synthesized without any surfactant assistance, the coating particles fell off easily after washing, which might have been caused by the weak attaching force (coordinated bond between ZnO and linen) induced by the deteriorated crystallinity.

The SEM image of functionalized fabric support show very clearly the individual yarns, having diameters of about 10-20 μm, covered by various ZnO aggregates (Fig.4). MCT- -CD can form complexes with divalent metal ions, due to its high number of coordinating functional groups (hydroxyl and glucoside groups) [31]. There is a possibility that the majority of the zinc ions were closely associated with the MCT- -

CD molecules. Based on the previous research, it can be claimed that nucleation and initial crystal growth of ZnO may preferentially occur on MCT- -CD [32]. Moreover, as polysaccharide, MCT- -CD showed interesting dynamic supramolecular associations facilitated by inter- and intra-molecular hydrogen bonding, which could act as matrices for nanoparticle growth in size of about 30–40 nm. They aggregated to irregular ZnO–CMC nanoparticles in a further step.

Table 2: Surface composition from EDX measurements at sample 6

Element	Wt%	At%
CK	25.34	46.47
NK	15.45	24.30
OK	08.91	12.27
BrL	01.90	00.52
CaK	00.63	00.35
ZnK	47.77	16.10
Matrix	Correction	ZAF

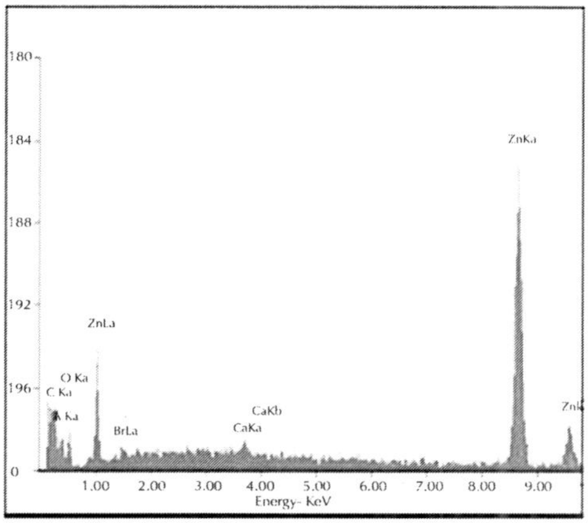

Figure 5: EDX analysis (sample 6) (Wt.: weight percent, at: atomic percent). [10].

The outcome of the EDX elemental analysis for sample 6 illustrated in Figure 5 and Table 2, show that surface composition contain approximately 47% ZnO, meaning that ZnO phase represented almost half of the sample mass.

The X-ray diffraction patterns of samples 4-7 compared with reference 2 are represented in Fig. 6:

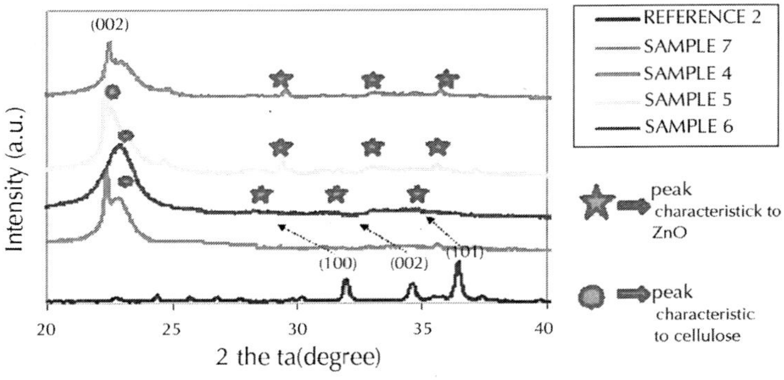

Figure 6: (Color online) XRD patterns of: Reference 2; Sample 4; Sample 5; Sample 6; Sample 7[10] The arrows indicate the peaks shift, in terms of working conditions. The height of the peaks has been multiplied by a 40 factor.

Figure 6 shows the selected-area diffraction pattern (2 =20-40°) of the obtained samples. The obtained XRD pattern and indexed lines of ZnO (reference 2) are presented in Figure 6. According to the literature [33], all the diffraction lines are assigned to the wurtzite hexagonal phase structure.

The composites patterns (sample 3-6) reveal both the presence of the peaks positions that matched well with those of the ZnO XRD pattern - lines (100), (002) and (101) - and the main peak of cellulose - linen (002) [34]. The small relative intensity of the peaks of the ZnO–linen composites is not well correlated with the EDX analysis, which showed a high content of deposited ZnO. The observed ZnO diffraction lines shift (samples 4 and 5) denotes the fact that the growth of the ZnO takes place not only on the support surface, but also inside the nanocavities due to the fibers roughness.

The intensities of the diffraction peaks decrease when the synthesis takes place with the assistance of the surfactant that prevent crystal growth in these working conditions (Fig.6).

In Fig. 7, the *FTIR spectrum* of hydrothermally synthesized, non-calcinated ZnO powder exhibited a high intensity broad band at about 430 cm^{-1} due to the stretching of the zinc and oxygen bond.

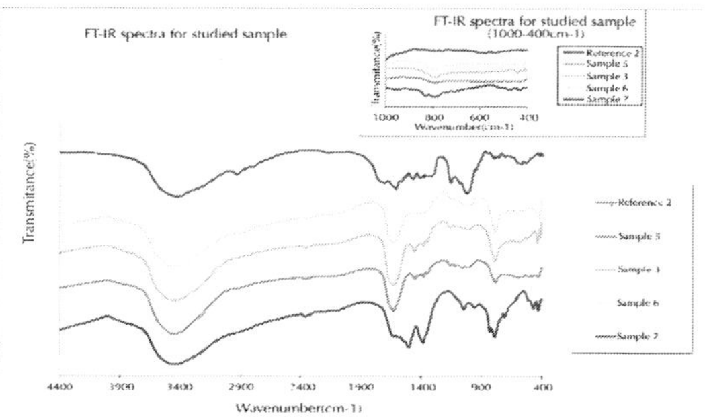

Figure 7: a) FTIR of spectra of: sample 3; sample 5; Sample 6; sample 7; Reference 2 [10].

As shown in the FTIR spectrum of MCT- -CD, the absorption bands between 1000 and 1200 cm^{-1}were characteristic of the – C –O– stretching on polysaccharide skeleton. A similar band was also observed in synthesized ZnO composites. And two peaks appeared at 1420 and 1610 cm^{-1}corresponding to the symmetrical and asymmetrical stretching vibrations of the carboxylate groups [35]. The peak at 2920 cm^{-1} was ascribed to C–H stretching associated with the ring methane hydrogen atoms. A broad band centered at 3450 cm^{-1} was attributed to a wide distribution of hydrogen-bonded hydroxyl groups. The FTIR spectra indicated that in ZnO–MCT- -CD nanoparticles, there was the strong interaction, but no obvious formation of covalent bonds between MCT- -CD) and ZnO.

Water vapors sorption behavior. Isothermal studies can be performed as a function of humidity (0-95%) in the temperature range 5° C to 85° C, with an accuracy of ± 1% for 0 - 90% RH and ± 2% for 90 - 95% RH. The relative humidity (RH) is controlled by wet and dry nitrogen flows around the sample. The RH is held constant until equilibrium or until a given time is exceeded, before changing the RH to the next level.

The vapours pressure in the sample room has been achieved by 10 steps of 10% humidity, each of them having a time of equilibrium setting between 10-20 minutes. At each phase, the weight adsorbed by the sample is measured by electromagnetic compensation between tare and sample, when the equilibrium is reached. Apparatus has an anti-condensation system for the cases that vapors pressure is very close/near to that of saturation. The cycle is finished by decreasing in steps of vapors pressure, in order to obtain desorption isotherms, as well.

Prior to measuring of sorption-desorption isotherms, drying of the samples is performed in nitrogen flow (250 mL/min) at 25°C, until the sample weight reached a constant value, at a relative humidity less than 1%.

The sorption/desorption isotherms recorded in these circumstances are shown in Fig.8.

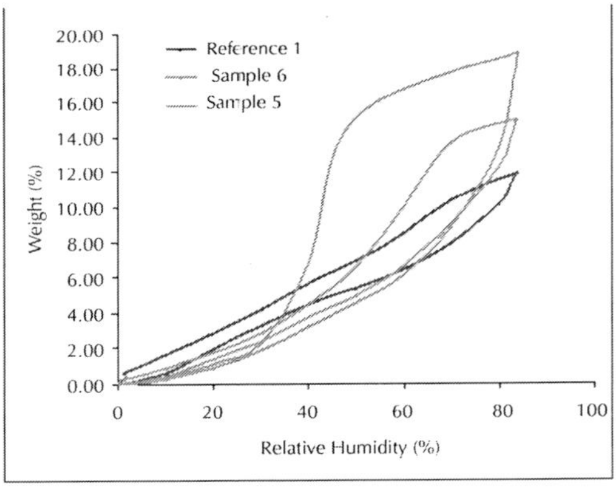

Reference 1; Sample 5; Sample 6 [10]

Figure 8: Comparative plots of rapid isotherms for water vapors sorption for the studied samples.

The reference sample (the linen fibrous support – yarn - unfunctionalized) has a smaller sorption capacity compared to that of

Sample 6 and Sample 5. High values of water vapors sorption capacity for the two last samples prove the fact that the material surface becomes more hydrophilic, more porous, respectively as it could be observed from hysteresis shape.

One of the main objectives of this review was to stress the adsorptive attributes, taking into account the improving of ZnO synthesis conditions. Consequently, the role of P123 in the ZnO synthesis was to obtain a composite with a higher porosity, in order to achieve the surface hydrophilicity, since there is direct correlation between porosity and hydrophilicity [36].

The shape of the moisture sorption isotherms for those two compounds is similar to those characteristic of mesoporous materials (type IV, according to IUPAC classification – with low sorption at low water vapor sorption (adsorption/desorption), moderate sorption at average humidity and rapidly increasing water sorption at high humidity). This type of isotherm describes the sorption behavior of hydrophilic material [37]. When a material is exposed to environmental water vapors, the water molecules firstly react with surface polar groups and form a molecular monolayer.

Based on the sorption studies, the IGAsorp software allows an evaluation of both monolayer and surface area value, by using BET (Brunauer-Emmett-Teller) model (Tabel 2).

Table 3: The main parameters of (water vapors) sorption-desorption isotherms for the studied samples

Sample	**Sorption capacity (%d.b.)**	**BET analysis**	
		A_{BET} (m^2/g)	**Monolayer (g/g)**
Reference 1	11.89	157.010	0.044
Sample 6	14.93	213.99	0.060
Sample 5	18.89	321.39	0.091

BET (1) equation is very often used for modeling of the sorption isotherms:

$$W=Wm\cdot C\cdot RH\ (1-RH)\cdot(1-RH+C\cdot RH) \quad (2)$$

where:

W- the weight of adsorbed water, Wm- the weight of water forming a monolayer, C – the sorption constant, *p/po=RH*- the relative humidity.

The sorption isotherms described by BET model up to a relative humidity of 40% are in relation to the sorption isotherm and material type. This method is mainly limited for II type isotherms, but can describe the isotherms of I, III and IV type [38-40], as well. The increasing water sorption is reflected both by the augmentation of monolayer and surface area values calculated with BET model (Tabel 3).

In Figure 8 the kinetic curves for humidity (water vapors) sorption/desorption processes for two of the samples are displayed. It is noticed that the time necessary for equilibrium setting for sorption processes is bigger than that of desorption. Sorption rate is smaller than that of desorption.

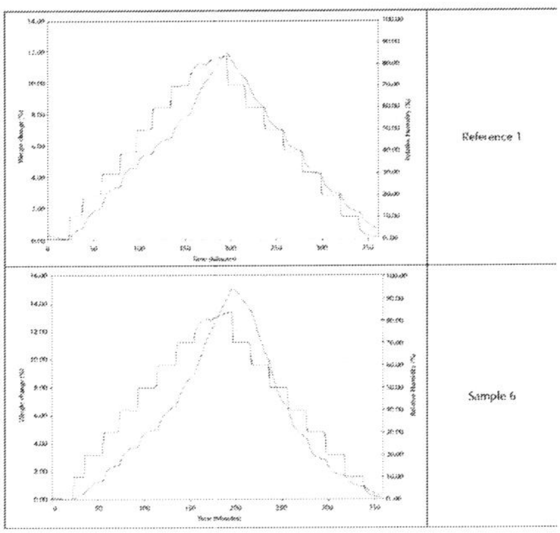

Reference 1; Sample 6 [10]

Figure 9: Kinetic curves for sorption/desorption processes of water vapors in the studied samples.

In Table 4 the dynamic moisture sorption capacity calculation was made using the equation written below, after the samples was kept at RH=90%, until the mass became constant:

Sorption capacity at RH=90% (%) $=\frac{W_{RH=90}-W_{RH=0}}{W_{RH=0}}\cdot 100$

As can be observed the obtained values are larger than those in the isotherms, this demonstrates the time necessary for reaching the equilibrium sorption is longer.

Table 4: Water vapor sorption capacity and speed for a longer time (until sample weight remains constant at a relative humidity of 90%)

Sample	**Weight at RH=0% (mg)**	**Weight at RH=90% (mg)**	**Sorption dynamic capacity RH=90% (%)**	**Time (s)**	**Sorption rate (10^{-3}%/s)**
Reference 1	4.58	5.14	12.28	32	3.82
Sample 6	5.32	6.34	19.14	50	3.75
Sample 5	5.56	6.83	22.77	40	5.58

In case of sample 5, the DVS analysis were made at two temperatures (25 °C and 35 °C), and the influence of this parameter on the sorption/desorption isotherms and kinetics are presented in Figure 8and Figure 9 respectively.

From Table 5, it is noticeable the augmentation of temperature conducts to an increase on vapor sorption capacity of the sample (probably due to the hydrogen bonds formation favoring sorption).

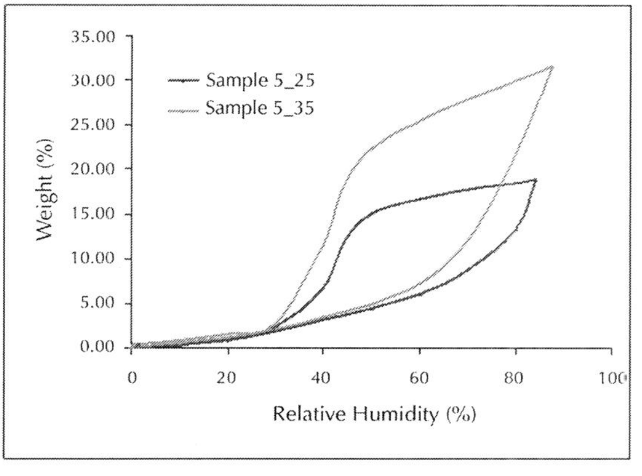

Figure 10: Comparative plots of rapid water vapors sorption/desorption isotherms for sample 5 [10] at both 25 °C and 35 °C respectively.

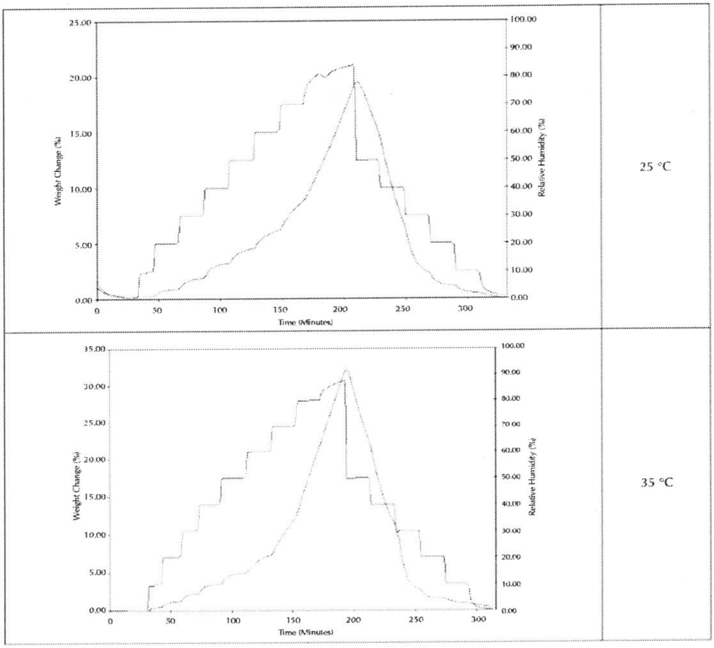

Figure 11: Kinetic curves for sorption/desorption processes of water vapors for sample 5 at both 25 °C and 35 °C respectively [10].

Table 5: Water vapor sorption capacity for sample 5 at both 25 °C and 35 °C respectively

Sample	**Sorption capacity (%)**
Sample 5_25	18.89
Sample 5_35	31.59

The differences between sorption-desorption speeds of those two temperatures for Sample 5 indexed sample are clearly highlighted by the presence of modified hysteresis and also by the kinetics curves

Thermal Degradation Mechanism of Linen Fibrous Supports Treated With ZnO

Considerable attention has been devoted to complete or correlate the results provided by the XRD analysis, with the DSC studies, since the last type of investigation is able to evaluate the crystallization/melting processes.

Vrinceanu et al tested thermal attributes of fibrous supports - ZnO nanocomposites under nitrogen [41]

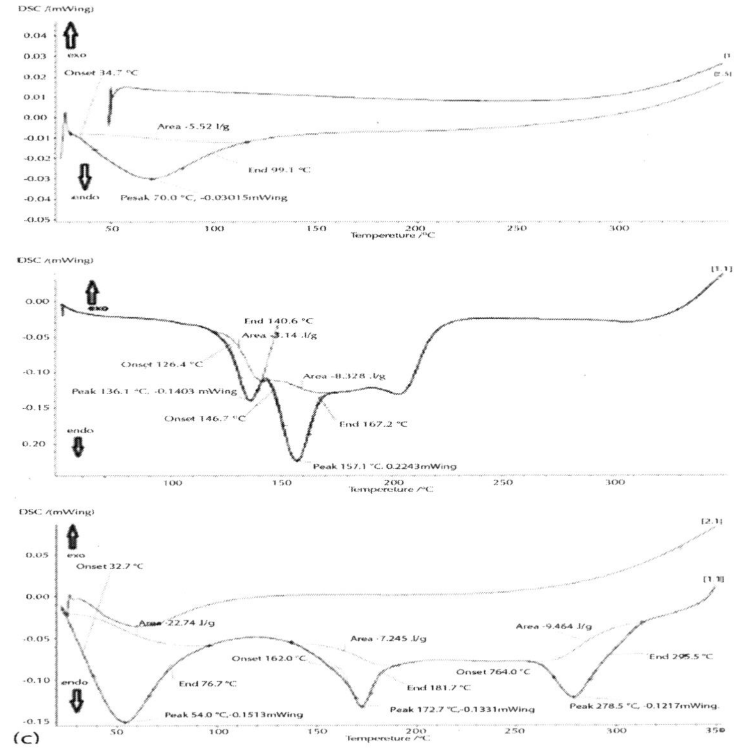

Figure 12: Typical DSC curve under nitrogen for: a Sample 3; b Sample 4; c. Sample 5 [41].

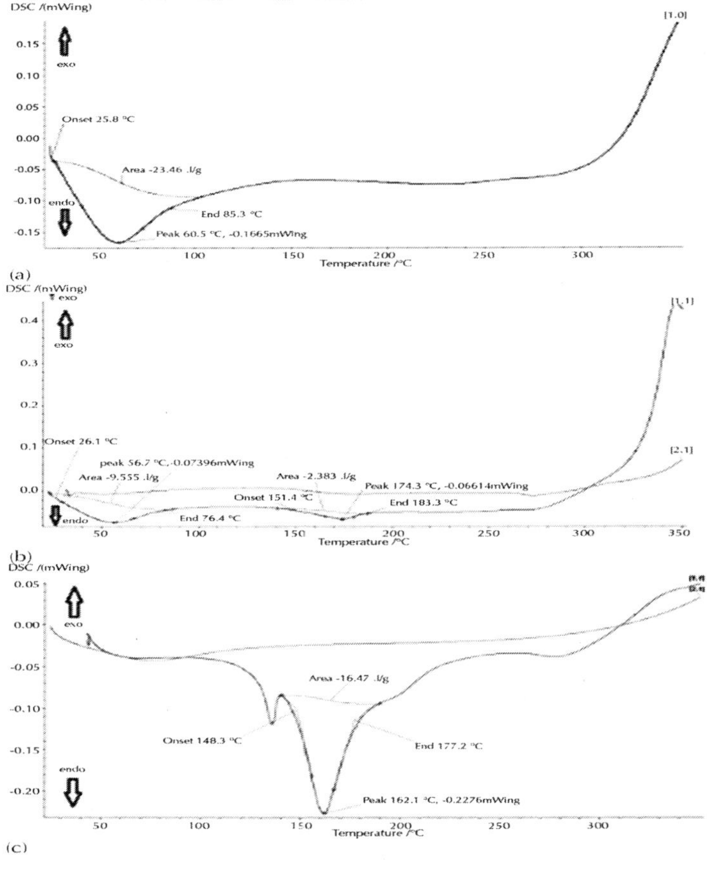

Figure 13: Typical DSC curve under nitrogen for: a – Sample 4; b – Sample 5; c – Sample 6 [41].

In the range 370°-395°C, in a typical DSC curve of cellulosic fibres, there is an endothermic peak, which has been shown to be primarily due to the production of laevoglucosan [42].

For linen fibres, this peak is sometimes partly or completely marked by an exothermal effect around 340°C, attributed to a base-catalysed-dehydration reaction that takes place in the presence of alkaline ions, such as those of sodium [43].

From 200 to 250°C a progressive mass loss associated with water release was observed. From the literature it is well known that lignocellulosic fibers degrade in several steps; the cellulose degrades between 310°–360°C, whereas the hemicellulose degrades at about 240°–310°C, and the lignin has been shown to degrade in wide temperature interval (200°–550°C) [44]. Technically speaking, it is not possible to separate the different degradation processes of the fiber components because the reactions are very complex and overlap in the range of 220°–360°C. It is noteworthy that the nanocomposite treated with ZnO nanoparticles with the assistance of MCT started to decompose at higher temperature than sample treated in the same conditions but without the presence of zinc oxide. Nevertheless, the existence of the MCT on the surface of the probes delayed the thermal degradation of the fibrous linen samples, even the non-treated with the zinc oxide particles.

It can be claimed that cellulose is thermally decomposed through two types of reactions. At lower temperatures, there is a complex process of gradual degradation including dehydration, depolymerisation, oxidation, evolution of carbon monoxide and carbon dioxide, and formation of carbonyl and carboxyl groups, ultimately resulting in a carbonaceous residue forms.

The endothermic band around 260°C from DSC curves (Fig. 14 (a) and (b)) indicates a weight loss. The surface acidity of zinc oxide nanoparticles keeps accelerating the decomposition of the fibrous substrate, as the temperature rises to 310°C, According to the FTIR spectra, a very much lower amount of carbonyl groups is found in the linen - ZnO nanocomposite specimens.

Meanwhile, MCT having a higher thermal conductivity as well as a greater heat capacity value absorbs the heat transmitted from the surroundings and retard the direct thermal impact to the polymer backbone [45, 46]. As a consequence, zinc oxide stabilizes the polymer molecules of the underneath substrates and delays the occurrence of major cracking up to 400°C (Fig. 15).

The masking effect of an exothermal reaction on the endothermic cellulose decomposition was clearly highlighted by the behavior of the reference fibrous linen (non-functionalized) subjected to the thermal treatment in N_2; it shows an exothermal peak at 260°C with a decreased enthalpy after the thermal treatment; the exothermal effect is attributable to -cellulose decomposition as observed in a curve of a cotton sample. Surprisingly, even within the second cycle of thermal treatment, the sample exhibits a similar exothermal peak at 363°C.

SUMMARY AND OUTLOOK

The review has been focused on a series of MCT- -CD grafted linen fibrous support (yarn) in whose matrices zinc oxide nanoparticles have been introduced with the assistance of two different surfactants. The coating particles fell off easily for the ZnO powder hydrothermally synthesized without any surfactant assistance after washing, which might have been caused by the weak attaching force (coordinated bond between ZnO and linen) induced by the deteriorated crystallinity.

Wetting characteristics are influenced by the type of surfactant used during the hydrothermal synthesis. It is in direct implication onto the relationship between the morphological, structural and chemical attributes and water vapor sorption-desorption behavior. Hydrophobicity of these fibrous composites has increased and based on the sorption/desorption isotherms registered by DSV, BET surface area, as well as XRD measurements were estimated, and assimilated these fibrous composites, set by IUPAC, with mesoporous materials. Humidity loss and drying speed of water from these studied samples depend of the type of surfactant.) A quantification of samples, in terms of their thermal stability has been surveyed, as well. Thus, this paper review intends to develop an innovative and more appropriate synthetic procedure and characterization of Nano scale ZnO coated fibrous composites under favorable conditions, by using the synergic effect of MCT and CTAB/P123 as surfactants. Prominent assessed attributes were emphasized:

- thermal stability and degradation mechanism of ZnO Nano coated linen fibrous samples;
- Cumulative barrier attributes conferred by the new components that interfered in the preparation technique: CTAB/P123 and MCT

These new features are believed to be the promising new lines of exploration of Nano scale ZnO coated fibrous composites in textile area.

ACKNOWLEDGEMENTS

The authors would like to greatly acknowledge the financial support provided by the two research contracts: /89/1.5/S/49944 POSDRU Project and PN-II-RU-TE-2011-3-0038 project, belonging to "Al.I.Cuza" University of Iasi.

REFERENCES

1. Weber, J., Futterer, C., Gowri, V.S., Attia, R., Viovy, J.L., La Houille Blanche 5:40 (2006); Saxana, M, Gowri. V.S., J Polym Compd 24:428 (2003); Gowri, V.S., Šaxena, MJ Chem Technol 14:145 (1997)
2. Raveendran, P., Fu, J., Wallen, S.L., Completely "green" synthesis and stabilization of metal nanoparticles, J. Am. Chem. Soc. 125, 13940–13941 (2003)
3. Mucalo, M.R., Bullen, C.R., Arabinogalactan from the Western larch tree: a new, purified and highly water-soluble polysaccharide-based protecting agent for maintaining precious metal nanoparticles in colloidal suspension, J. Mater.Sci. 37, 493–504 (2002)
4. He, J.H., Kunitake, T., Nakao, A., Facile in situ synthesis of noble metal nanoparticles in porous cellulose fibers, Chem. Mater. 15, 4401–4406 (2003)
5. He, J.H., Kunitake, T., Nakao, A., Facile fabrication of composites of platinum nanoparticles and amorphous carbon films by catalyzed carbonization of cellulose fibers, Chem. Commun. 4, 410–411 (2004)
6. Walsh, D., Arcelli, L., Ikoma, T., Tanaka, J., Mann, S., Dextran templating for the synthesis of metallic and metal oxide sponges, Nat. Mater. 2, 386–390 (2003)
7. Vigneshwaran, N., Kumar, S., Kathe, A.A., Varadarajan, P.V., Prasad, V., Functional finishing of cotton fabrics using zinc oxide–soluble starch nanocomposites, Nanotechnology 17, 5087–5095 (2006)

8. Ma, X.F., Chang, P.R., Yang, J.W., Yu, J.G., Preparation and properties of glycerol plasticized-pea starch/zinc oxide–starch bionanocomposites, Carbohydr. Polym. 75, 472–478 (2009)
9. Radhakrishnan, T., Georges, M.K., Nair, P.S., 2007. Study of sago starch–CdS nanocomposite films: Fabrication, structure, optical and thermal properties, J.Nanosci. Nanotechnol. 7, 986–993 (2007)
10. Diana Tanasa, Narcisa Vrinceanu, Alexandra Nistor,Claudia Mihaela Hristodor, Eveline Popovici, Ionut Lucian Bistricianu, Florin Brinza1, Daniela-Lucia Chicet,Diana Coman, Aurel Pui, Ana Maria Grigoriu, Gianina Broasca, Zinc oxide-linen fibrous composites: nmorphological, structural, chemical and humidity adsorptive attributes, Textile Research Journal, 82(8) 832–844 (2012)
11. Chandramouleeswaran, S., Mhaskel, S., Kathe, A.A., Varadarajan, P.V., Prasad, V., Vigneshwaran, N., Functional behaviour of polypropylene/ZnO–soluble starch nanocomposites, Nanotechnology 18, 385702 (2007)
12. Tang Z K, Wong G K L, Yu P, Kawasaki M, Ohtomo A, Koinuma H and Segawa Y Appl. Phys. Lett. 72 3270 (1998)
13. CorralesT, Catalina F, Peinado C, Allen NS, Fontan E. Photooxidative and thermal degradation of polyethylenes interrelationship by chemiluminescence, thermal gravimetric analysis and FTIR data. J Photochem Photobiol. A 2002;147:213-24
14. Gawas UB, Verenkar VMS, Mojumdar SC. Synthesis and characterization of Co0.8Zn0.2Fe2O4 nanoparticles. J Therm Anal Calorim. 2011; 104:879-883.
15. Gonsalves LR, Mojumdar SC, Verenkar VMS. Synthesis and characterization of Co0.8Zn0.2Fe2O4 nanoparticles. J Therm Anal Calorim. 2011;104:869-873
16. Gonsalves LR, Mojumdar SC, Verenkar VMS. Synthesis of cobalt nickel ferrite nanoparticles via autocatalytic decomposition of the precursor. J Therm Anal Calorim. 2010; 100:789-792.
17. Gawas UB, Mojumdar SC, Verenkar VMS. Synthesis of cobalt nickel ferrite nanoparticles via autocatalytic decomposition of the precursor. J Therm Anal Calorim. 2010; 100:867-871.

18. Verdu J, Rychly J, Audouin L. Polym Degrad Stab. 2003; 79:503-9.
19. Allen NS, Edge M, Corrales T, Childs A, Liauw CM, Catalina F, et al. Ageing and stabilization of filled polymers: an overview. Polym Degrad Stab. 1998; 61:183-99.
20. Mojumdar SC, Moresoli C, Simon LC. Legge RL. Edible wheat gluten (WG) protein films: Preparation, thermal, mechanical and spectral properties. J Therm Anal Calorim. 2011; 104:929-936.
21. Gawas UB, Mojumdar SC, Verenkar VMS. J Therm Anal Calorim. 2009;96:49-52
22. Mocanu AM, Odochian L, Apostolescu N, TG-FTIR study on thermal degradation in air of some new diazoaminoderivatives. J Thermal Anal Calorim. 2010; 100 (2):615-622.
23. Singhal M, Chhabra V, Kang P, Shah DO. Mater Res Bull. 1997; 32:239-247
24. Hingorani S, Pillai V, Kumar P, Multani MS, Shah DO. Mater Res Bull. 1993; 28:1303-1310.
25. Fan Q, John J, Ugbolue SC, Wilson AR, Dar YS, Yang Y. AATCC Rev. 2003; 3(6):25.
26. Grigoriu A.M., Cercet ri în domeniul compusilor de incluziune ai ciclodextrinelor si al derivatilor acestora cu aplicatii în industria textil . Ph. D. Diss., Iasi (2009)
27. Reuscher H, Hinsenkorn R. BETA W7 MCT-new ways in surface modification. J Incl Phenom Macrocycl. 1996; 25:191–196.
28. Ogata N, Ogawa T, Ida T, Yanagawa T, Ogihara T, Yamashita A. Sen'i Gakkaishi. 1995; 51(9), 439.
29. Patterson A. The Scherrer Formula for I-Ray Particle Size Determination. Phys Rev. 1939; 56 (10): 978–982.
30. Xiao-Juan J, Pascal Kamdem D. Chemical composition crystallinity and crystallite cellulose size in populus hybrids and aspen. Cellul Chem Technol. 2009;43(7-8):229-234
31. Reuscher H., Hinsenkorn R., Journal of inclusion phenomena and macrocyclic chemistry, 25, 191–196 (1996)
32. Taubert, A., Wegner, G., Formation of uniform and monodisperse zincite crystals in the presence of soluble starch. J. Mater. Chem. 12, 805–807 (2002)

33. J. Yu et al. Bioresource Technology 100 2832–2841 (2009)

34. Sunkyu Park1,3, John O Baker1, Michael E Himmel1, Philip A Parilla and David K Johnson, Cellu lose crystallinity index: measurement techniques and their impact on interpreting cellulase performance, Biotechnology for Biofuels 2010, 3:10 Jiugao, Yu, Jingwen, Yang. Baoxiang, Liu, Xiaofei, Ma, Preparation and characterization of glycerol plasticized-pea starch/ZnO–carboxymethylcellulose sodium nanocomposites, Bioresource Technology 100, 2832–2841 (2009)

35. Gu, F., Wang, S.F., Lu, M.K., Zhou, G.J., Xu, D., Yuan, D.R. Langmuir, 20: 3528 (2004)

36. http://test.ttri.org.tw/neweng/RD/images/pub3.pdf

37. Brunauer, S., Deming, L.S., Deming, W.E., Teller, E. On a Theory of the van der Waals adsorption of gases, J Am Chem Soc, 62(7): 1723-1732 (1940

38. Guggenheim, E. A. Application of Statistical Mechanics, Clarendon Press, Oxford, 186-206 (1966).

39. Anderson, R.B. Modifications of the Brunauer, Emmett and Teller Equation, J Am Chem Soc, 68(4): 686-691 (1946)

40. De Boer, J.H. The Dynamical Character of Adsorption, 2nd ed., Clarendon Press, Oxford, 200-219 (1968).

41. Vrinceanu , N., Tanasa, D., Hristodor C.M., Brinza, F., Popovici E., Gherca D., Pui A., Coman, D., Carsmariu A., Bistricianu I., Broasca, G., Synthesis and characterization of zinc oxide nanoparticles Application to textiles as thermal barriers, J Therm Anal Calorim, DOI 10.1007/s10973-012-2269-7 (2012)

42. Ye DY, Farriol X. Preparation and characterization of methylcellulose from Miscanthus Sinensis. Cellul. 2005; 12, 507.

43. Revola JF, Dietricha ND D, Goring AI. Effect of mercerization on the crystallite size and crystallinity index in cellulose from different sources. Can J Chem. 1987; 65, 1724.

44. [44].Nevell TP, Zeronian S H. Cellulose Chemistry and Its Applications M. Chichester: Ellis Horwood Ltd., 1985; 423−454.

45. Lee HJ, Yeo SY, Jeong SH. Antibacterial Effect of Nanosized Silver Colloidal Solution on Textile Fabrics. J Mater Sci. 2003; 38:2199–2204.

46. Riva A, Algaba IM, Montserrat P. Action of a finishing product in the improvement of the ultraviolet protection provided by cotton fabrics. Modelisation of the effect. Cellul. 2006; 13:697–704.

Citations

CHAPTER 1

Se Woon Choi, EunMi Kwon, Yousok Kim , Kappyo Hong, and Hyo Seon Park, A Practical Data Recovery Technique for Long-Term Strain Monitoring of Mega Columns during Construction, doi:10.3390/s130810931.

CHAPTER 2

Lei Li, Qingbin Li and Fan Zhang (2013). Applications of SMA Bundles in Practical Concrete Structures, Shape Memory Alloys - Processing, Characterization and Applications, Dr. Francisco Manuel Braz Fernandes (Ed.), ISBN: 978-953-51-1084-2, InTech, DOI: 10.5772/51207.

CHAPTER 3

Pengfei Liu, Qilin Zhao, Fei Li, Jinchun Liu, and Haosen Chen, Research on the Mechanical Properties of a Glass Fiber Reinforced Polymer-Steel Combined Truss Structure, doi: 10.1155/2014/309872.

CHAPTER 4

V. Susila Anand, C. Kavitha, and C. V. Subbarao, Effect of Cavity Design on the Strength of Direct Posterior Composite Restorations: An Empirical and FEM Analysis, doi: 10.1155/2011/214751.

CHAPTER 5

Libin Wang, Hui Jin,,Haiwei Dong, and Jing Li, Balance Fatigue Design of Cast Steel Nodes in Tubular Steel Structures, doi: 10.1155/2013/421410.

CHAPTER 6

Mohammad Panjehpour, Hwa Kian Chai, and Yen Lei Voo, Strut Deformation in CFRP-Strengthened Reinforced Concrete Deep Beams, doi: 10.1155/2014/265879.

CHAPTER 7

Diego Manfredi, Flaviana Calignano, Manickavasagam Krishnan, Riccardo Canali, Elisa Paola Ambrosio, Sara Biamino, Daniele Ugues, Matteo Pavese, and Paolo Fino (2014). Additive Manufacturing of Al Alloys and Aluminium Matrix Composites (AMCs), Light Metal Alloys Applications, Dr. Waldemar A. Monteiro (Ed.), ISBN: 978-953-51-1588-5, InTech, DOI: 10.5772/58534.

CHAPTER 8

Narcisa Vrinceanu, Alina Brindusa Petre, Claudia Mihaela Hristodor, Eveline Popovici, Aurel Pui, Diana Coman, and Diana Tanasa (2013), Zinc Oxide — Linen Fibrous Composites: Morphological, Structural, Chemical, Humidity Adsorptive and Thermal Barrier Attributes, Modern Surface Engineering Treatments, Dr. M. Aliofkhazraei (Ed.), ISBN: 978-953-51-1149-8, InTech, DOI: 10.5772/55705.

Index

A

B

C

L

M

N

O

P

R

S

T

U

W

X